AF526821

Cultivation and Uses of Aromatic Plants

Ramesh Kumar Srivastava
Shakti Vinay Shukla
Sanjeet Singh Dagar
Fragrance & Flavour Development Centre (FFDC)
(Ministry of MSME, Govt. of India)
Kannauj (U.P.)

ibdc publishers

Published by

ibdc publishers

Khushnuma Complex Basement
7, Meerabai Marg (Behind Jawahar Bhawan),
Lucknow 226 001 U.P. (INDIA)
Tel. : 91-522-2209542, 2209543, 2209544, 2209545
Fax : 0522-4045308
E-Mail : ibdco@airtelmail.in

First Edition 2009

Price: Rs

ISBN 978-81-8189-479-3

© Publisher
All Rights Reserved

No part of this publication may be reproduced, stored in a retrieval system, or transmitted, in any form or by any means, electronic, mechanical, photocopying, recording or otherwise, without the prior written permission of the publisher.

Composed & Designed at :

Panacea Computers
3rd Floor, Agarwal Sabha Bhawan, Subhash Mohal
Sadar Cantt. Lucknow-226 002
Phone : 0522-2483312, 9335927082, 9452295008
E-mail : prasgupt@rediffmail.com

Printed at:

Salasar Imaging Systems
C-7/5, Lawrence Road Industrial Area
Delhi - 110 035
Tel. : 011-27185653, 9810064311

माधव लाल
अपर सचिव एवम्
विकास आयुक्त
MADHAV LAL
Additional Secretary &
Development Commissioner

सत्यमेव जयते
MSME
MICRO, SMALL & MEDIUM ENTERPRISES

भारत सरकार
सूक्ष्म, लघु एवं मध्यम उद्यम मन्त्रालय
GOVERNMENT OF INDIA
MINISTRY OF MICRO,
SMALL AND MEDIUM ENTERPRISES
'A' WING 7TH FLOOR, NIRMAN BHAVAN,
NEW DELHI-110108
Phones: 23061176, Fax: 23062315

Foreword

I am delighted to write the foreword for this book - ***Cultivation and Uses of Aromatic Plants,*** which will be useful for farmers, entrepreneurs, students, traders and exporters of this country, who are related to the aroma industry.

In the context of diversification in agriculture, the aromatic plants offer excellent alternative possibilities. With diverse agro-climatic conditions, India is emerging as a major producer, consumer and exporter of aromatic plants and their essential oils in the world. To capitalise on this potential, a holistic knowledge, comprising of cultivation techniques, processing, standardisation, quality assessment and marketing is required. The Fragrance & Flavour Development Centre (FFDC), Kannauj (under the Ministry of Micro, Small & Medium Enterprises) is a unique institution adopting a holistic approach in this sector since 1992.

Although a number of research papers and other documents are available on each of these subjects, a handy and comprehensive book, which could be easily used by different stakeholders, has been missing. This book, which delves into aspects of cultivation and processing of 15 important aromatic crops (classified according to agroclimatic conditions), distillation technologies and details of agencies which could assist entrepreneurs in this field, will be very beneficial.

I congratulate the authors for putting together this book and hope that similar initiatives will help in growth of fragrance and flavour industry in the country.

6/8/09

(Madhav Lal)

Preface

The Indian sub-continent blessed with the world's richest resources of aromatic plant and their uses since time immemorial in different form by the collection from wild sources. India comprises its botanical wealth is about 45,000 plant species including 18,500 flowering and essential oil bearing plants. The aromatic plants are the main source of essential oils which is complex mixture of naturally occurring odoriferous compounds that exists in form of oil glands of flowers,leaves, woods, barks, roots and seeds. Chemically essential oils are hydrocarbons and oxygenated hydrocarbons which can be grouped into terpenes, esters, alcohols, aldehydes, ketones, phenols, etc. In other words, essential oils are volatile extracts of plants used as base material in perfumes (fragrances and flavours), cosmetics, aromatherapy, attars and aromatherapy.

Presently, fragrant raw materials are collected from wild and well as cultivated sources for completing the demand of the perfumery and cosmetic industry. This demand in organized sector of industry has led to their introduction into agriculture which besides meeting the demand and reasonable price has also enable the produces to maintain standards on yield, quality, potency and chemical constituents of the produce. A large number of aromatic plants have thus come under systematic research efforts through multi-disciplinary investigations in India since independence. This activity has generated a wealth of valuable information on several facets of agricultural productivities like crop improvement, introduction of improved agro-technologies, extraction methodology, faster instrumental analysis, product development and multifarious uses of essential oils and their produce. This material is widely scattered in research journals, proceedings of seminars and symposia, technical bulletins, monographs and technology papers. We have made an efforts to bring all this information together in the form of a book "Cultivation and Uses of Aromatic Plants" for associated farmers, entrepreneurs, students, researchers, traders, exporters, user industries and libraries.

In day-to-day health care needs undisputed role of aromatic plants and its essential oils is of utmost important and these have to cultivated and propagated, because it has necessitated to reduce the pressure on natural resources, as forests, till date, continue to provide major share of raw material to the perfumery, pharmaceutical and other related industries and to check the use of spurious substitutes and adulterants, which have resulted in the deterioration of the standard of the products and preparations. The systematic and large scale cultivation and processing is only solution for providing genuine raw material for associated industries and minimize the over-exploitation of our genetic resources from forests.

The present book deals with improved cultivation and post-harvest technologies of 15 major aromatic plants, distillation technologies and their diverse uses based on experimental work and available authentic literatures. The colour photographs of aromatic crops have been soot-out in FFDC herbal garden and experimental fields for proper identification. We will gratefully welcome any useful suggestions regarding this book.

Date: 07.07.2009
Place: Kannauj

Dr. R.K. Srivastava
S.V. Shukla
Dr. S.S. Dagar

Acknowledgement

The authors are extending deep sense of gratitude to Shri Madhav Lal, IAS, Additional Secretary & Development Commissioner (MSME) and Chairman, FFDC Kannauj for encouraging for writing a book on cultivation and utilization of aromatic plants.

We also thankful to Shri Gorakh Prasad, Principal Director, FFDC; Sri KN Dwivedi, Director (Project); Dr Aditya Sharma, Dy. Director, Agro-Tech Division other staff member of the Fragrance and Flavour Development Centre (FFDC) for their direct and indirect co-operation in writing this book as and when required.

We are thankful to Sri A.P. Singh, Documentation Officer and Shri Ramkesh, DTP Operator for time to time help in literature search and typing and Shri Kamlesh Kumar for helping in experiment on agro-technological work.

We wish to regard our sincere thanks to Prof. Sushil Kumar, Ex. Director, Central Institute of Medicinal and Aromatic Plants, Lucknow and Senior INSA Scientist, NIPGR, New Delhi; Dr. Anil Kumar Singh, Scientist-F and Head, Technology and Variety Development, CIMAP, Lucknow; Prof. H.B. Singh, Banaras Hindu University, Varanasi; Dr. Aparbal Singh, Ex. Scientist, CIMAP, Lucknow; Dr. Alok Lehri, Senoir Scientist, National Botanical Research Institute, Lucknow and Dr N.K. Singh, Reader, MLK (PG) College, Balrampur for guiding time to time during writing of this book.

We also express our special thanks to Shri Shailendra Jain, President, Essential Oil association of India, New Delhi; Shri Akhilesh Pathak; Munnalal & Sons, Kannauj; Dr. G.S Jaryal, Bhopal; Dr. S.R. Singh, NIAM, Jaipur; Dr. R.C. Uniyal, AMITY University, NOIDA; Shri Ramakant Harlalka, Sankhubaba International, Mumbai; Shri I.N. Sahai, NHB, Lucknow; Shri Yogesh Dubey, Vice President EOAI & FAFAI and Dr. Kausal Kumar, CSAUT, Kanpur for their contribution and valuable discussion on aromatic plants, their utilization and market aspects.

We thankful to our progressive farmers like Shri P.C. Pathak; Shri Krishna Prasad, Bihar; Shri Sanjai Kushwaha, M.P.; Shri Sanjai Tripathi and Anurag Katiyar, UP for sharing their real experience on cultivation, economics, yield and marketing of aromatic crops.

We are overwhelming with rejoiceto avail this rare opportunity to evince our profound sense of indebtedness to our wives Mrs. Shobha Dagar W/o Dr. S.S. Dagar, Dr. Jyotsna Shukla W/o S.V. Shukla and Dr. Suchi Srivastava W/o Dr. R.K. Srivastava for devoting their time and cooperation at the time of writing of this book.

Date: 07.07.2009
Place: Kannauj

Dr. R.K. Srivastava
S.V. Shukla
Dr. S.S. Dagar

About the Authors

Dr. Ramesh Kumar Srivastava did his M.Sc. in Botany from Dr. Ram Manohar Lohia Avadh University, Faizabad (UP) in 1995. After his M.Sc. he joined Central Institute of Medicinal and Aromatic Plants (CIMAP), Lucknow in 1996 as a Junior Research Fellow and did his Ph.D. on menthol mint production technology and popularization among the farmers of Uttar Pradesh. During Ph.D. programme, he got exposure on cultivation, processing of important aromatic plants and trained modern technologies on plant science. After completing his Ph. D. from CIMAP, he joined Deepak Fertilizers and Petrochemical Corporation Limited, Pune as Deputy Manager on 2000. In this organization he spent one year and worked on cereal and vegetable crop improvement. Presently, Dr. Srivastava is working as Extension Officer (Agro-Technology and Extension Services) at Fragrance and Flavour Development Centre, Kannauj since 2002. Apart from this book he published 5 technical bulletins, 31 papers in national and international journal, 10 papers presented in national and international seminars and 1 US patent on aromatic and medicinal plants. He may be contacted through rks_ffdc@rediffmail.com.

Shakti Vinay Shukla, presently, Dy. Director, FFDC, Kannauj was born on 24 September 1968 in Varanasi. He received his B.Sc. (Hons.) in Chemistry & M.Sc. (Organic Chemistry) from Banaras Hindu University, Varanasi in 1988 & 1990 respectively. Thereafter he served Harcourt Butler Technological Institute as Technical Assistant from 1990 to 1992, where he worked on post harvest technology for extraction and storage of mustard & sesame seed, oil and cake. He then joined Fragrance & Flavour Development Centre (FFDC), Kannauj in 1992 as Scientific Assistant. Currently he is Dy. Director (Fragrance & Flavour) & head Fragrance & Flavour Division at the same centre. During his work at FFDC he received training at India & at abroad in the development of fragrance & flavour. He was responsible to establish the fragrance & flavour division at the centre. While working at FFDC he developed more than 250 formulations on fragrance & flavour. A number of them have been very successfully accepted by the industry. He is also working on aromatherapy and developed a number of products. He is also a member of executive committee Essential Oil Association of India, New Delhi, Indian Science Congress of Indian Chemical Society, All India Association of Anesthetists of Indian Medicine. He has more than 20 papers on essential oil, fragrance, flavour & aromatherapy to his credit. He has handled a number of projects for DRDE, Gwalior and CCRAS, New Delhi as well. He has received Young Scientist award by Council of Science & Technology, Government of Uttar Pradesh and many other awards by different associations.

Dr. Sanjeet Singh Dagar is presently Principal Director, Fragrance and Flavour Development Centre, Kannauj since April, 2009. Before this organization, he was Deputy Director (Horticulture) in Central Public Works Department, Delhi. He did his M.Sc. in Horticulture from CCS University, Meerut in 1989 and completed his Ph.D. in Horticulture on Tuberose plant in 1993 from CCS University, Meerut. He has more than 15 years experience on horticultural and aromatic crops production and management. Dr Dagar has got international exposure on horticulture and park management related works from Japan and Singapore. Apart from this book he published 5 paper related to horticulture and aromatic plants. He is member and life member of different societies related to horticulture and aromatic plants.

CONTENTS

Chapter 1

Introduction

Since time immemorial, in our country, the technology for producing essential oils has been existence from as early as the 3rd century B.C. and today our country has become a destination for the essential oils, fragrance and flavour industry of the world. The increasing demand of natural essential oils in the associated industries, the future will witness our country's strong and dominating position in the world. However, the rapid expansion of the herbal industry in the world, it is absolutely not feasible to overlook and encourage prevailing situation where 60% of the industry demand are being met through wild sources. This should be gradually reduced and be replaced by the improved cultivation technology by the farmers. Initially, the cultivation of aromatic crops is expected to be slow because of farmer's apprehensions about the economics and returns. This phase required assured price, government support in terms of subsidies and buy back guarantee to the farming communities. In this direction, this book will play important role to aware the farming communities about cultivation, distillation, processing, economics, R&D and financing institutes and their norms, etc.

The growth of perfumery raw materials rising very fast in the world trade, it is estimated that the total world production of essential oils, aroma chemicals, fragrances and flavour were about 3600 US$ in 1978, which rose to 1.85 billion US$ in 1983 and reached 7.5 billion US$ in 1987. Now the market is about 15 billion US$ and expected to increase in 21st century is about 30 billion US$. The market share of India is increasing during last five years from 20-30%. The total estimated world production of essential oils is about 1,08,000 to 1,12,000 ton per year. The data indicated that we have to explore the possibilities of cultivation and processing of important essential oil bearing plant for sustainable supply of the raw material to the associated domestic industry and export to generate foreign exchange. India with its diverse agro-climatic conditions and ample availability of resource has very strong position to play a vital role in the production and supply of natural essential oils to the world market. Presently, we are still imports some of the very important essential oils, like rose, patchouli, geranium, rosemary, tuberose and jasmine to meet our internal demand of essential oil based industry. With little efforts by the farmers and scientists, many more essential oils can be produced in available various agro-climatic zone of our country.

After introduction of menthol mint in our country in 1954 from china, presently we are the world leader in production and trade of mint and allied product. This effort was realized and secured top position in mint oil production after rigorous research by the scientists of CSIR, agriculture universities and FFDC with the real interest of farming communities to adopt the cultivation and processing. We can try in other important aromatic crops like lemongrass, palmarosa, citronella, patchouli, basil, rose, rosemary, marigold, tuberose, jasmine, chamomile, lavender, vetiver and geranium for cultivation in specific agro-climatic condition with existing cropping pattern to secure top position in production of essential oils to meet our internal demand and export. We have good opportunities to explore the available resources in terms of manpower and agro-climate.

Chapter **2**

Criteria for Selection of Aromatic Crops according to their Agro-climatic Region

The Indian agro-climate has varied diversity to grow any crops either suitable for temperate or tropical condition. Although, India is an under developed country but Indian spices, aromatic, medicinal plants and condiments have traditionally contributed a significant role to the local and overseas market and also there is lot of chances to spread in this trade. More than 65% world's required raw material can be cultivated or collected form Indian Territory. Indian climatic conditions have variations and hence, world's all known climatic zones are available in India.

S. No.	Agro-climatic zone	Suitable Crops
1.	Western Himalayan Region	Basil, Eucalyptus, Menthol Mint, Costus, Hops, Jatamansi, Saffron, Pyrethrum, Juniper Barry, Colchicum, Chamomile, Rose, Deodar, Valerian, Aconitum, Lavender.
2.	Eastern Himalayan Region	Tea, Eucalyptus, Clove, Kewada, Vanilla, lemongrass, cumin, Hops, Costus, cardamom, Colchicum, Pine, Agra, Citronella, Vetiver, hoe, Myrrh, Orchids.

3.	Lower Gangetic Plains	Dill, Coriander, Kewada, Garlic, Aniseed, Galangal Cumin, Nagarmotha, Calamus, Castor.
4.	Middle Gangetic Plains	Eucalyptus, Geranium, Menthol mint, Coriander, dill, Davana, Costus, Rose, Hops, Celery, Aniseed, Cumin, Basil, Calamus Lemongrass and Palmarosa.
5.	Upper Gangetic Plains	Eucalyptus, Rose, basil, Dill, Lime Costus, Celery, Mentha Cumin, Calamus.
6.	Trans Gangetic Plains	Eucalyptus, rose, Celery, Hos, Lemon,Vetiver, mentha, tuberose, fennel, Carrot Seed, Coriander Calamus.
7.	Eastern plateau and Hills	Palmarosa, Lemongrass, Nagaramotha, Cumin, tuberose, Oil Palms, Fenugreek, Citronella, Kewada, Dill, Coriander, black Pepper, Castor.
8.	Central plateau & Hills	Palmarosa, Vetiver, Coriander, Cumin, Sandalwood, Basil, Castor, Calamus, Poppy, turmeric, kewada, Dill, Lemon. Oil Palms, Fenugreek, henna.
9.	Western Plateau & Hills	Palmarosa, lemon, Turmeric, Coriander, Jasmine, Oil palms, Davana, Vetiver, Orange, ginger, Sandalwood, Tuberose, Castor, Galangal.

10.	Southern Plateau & Hills	Sandalwood, Rose Tuberose, ginger, Lemongrass Fenugreek, Geranium, Eucalyptus, Jasmine, Turmeric, Eucalyptus, lemongrass, Palmarosa, Nagarmotha, Agar, Nutmeg.
11.	East Coast plains & Hills	Black pepper, Cinnamon, Ginger, tuberose, Citronella, Basil, Vetiver, Galangal, Clove. Turmeric, Eucalyptus, Lemongrass Palmarosa, Nagarmotha, Agar, Nutmeg.
12.	West Coast plains & Ghats	Black pepper, Cinnamon, Lemongrass, Palmarosa, Vetiver, Vanilla, Ginger, Nutmeg, Clove, Cardamom, Citronella, Nagarmotha, Sandalwood, Turmeric.
13.	Gujarat Plain & Hills	Psyllium Vetiver, Coriander, Kewada, Jojoba, Basil, Galangal, Poppy, Cumin, fenugreek, Lime, Aniseed, Caster, Henna.
14.	Western Dry Regions	Castor, Cumin, Psyllium, Jojoba, Henna, fenugreek, Poppy, Coriander, Rose.
15.	Island Zones	Eucalyptus, patchouli, Oil palm, Ylang Ylang, Clove, Kewada, Nutmeg, Citronella, Lemongrass Cinnamon.

Chapter **3**

Detail Cultivation Technology of Aromatic Crops

BASIL
(*Ocimum basilicum*)

Cultivation Technology of Basil

Sweet basil is botanically known as *Ocimum basilicum* L. belongs to family lamiaceae. About 150 species have been found in the world, out of which some are wildly grown in India. A few are commercially cultivated for their important essential oil and crude drug purpose but the Sweet Basil or French Basil is cultivated for its essential oil.

Apart from sweet basil, some of them are also cultivated for their chemical constituents but the sweet basil is widely cultivated in India as well as in other parts of the world. The sweet basil is native of India but also cultivated in France, Italy, Bulgaria, Egypt, Hungary, South Africa and USA. The oil of this species contains methyl chavicol, linalool and some other minor constituents. (Plate 1)

Ocimum basilicum is erect, herbaceous, strongly aromatic, annual plant having height of 75 to 95 cm. Stem often branched from the base, purplish and subglabrous in lower part, four angular, intensely hairy higher up. Leaves are 3 to 6 cm long and 1.5 to 3.5 cm in diameter, ovate, lanceolate with acuminate or rounded base, acute or subobtuse, subentire to serrate, thinly hairy and gland pinnate. Flowers are born in long terminal racemose inflorescences. Pedicles densely hairy, calyx 0.3 to 0.4 cm long at first, afterwards 0.5-0.6 cm long hairy outside, glabrous inside, at base with dense whorl of long hairs above it, glandulose upper tip ovate to rounded. Corolla white or pale purple tube, stamens exert, posterior filaments shorts, white hairy and transverse processes.

Production technology/cultivation

Indian basil is only *ocimum* species cultivated on large scale in Indian states. Mostly, it is cultivated for its important essential oil in the states of Uttar Pradesh, Punjab, Haryana, Madhya Pradesh and some other parts of India.

Soil and Climate

The crops grown well on moderate fertile and well drained sandy loam to loam soils with good water holding capacity with a pH ranges from 5 to 8. It requires moderate climate (Temp. 15^0 to 35^0C). North Indian plains are suitable for its cultivation. In hills, it can be cultivated only as a summer crop.

Preparation of land

The field should be disc ploughed once and harrowed twice with the mixing of 12-15 tonnes FYM per hectare. It should be ready for planting in June to first week of July.

Propagation

The crop can be propagated through nursery as well as direct sowing/broadcasting. However, direct sowing is not advisable for cultivation due to lower yield.

(I) Raising of nursery

The cultivation of Indian basil through raising of nursery is recommended. The nursery bed should be cleaned of stubbles, weeds and soil nematodes, 1.5 to 2.0 tonnes FYM, 20 kg each of Nitrogen, Phosphorous and Potash should be applied. About 500 m^2 nursery area would be required for one ha. plantation. The size of beds should be 2 x 4 meter with proper irrigation channel. As the seeds are small, it should be mixed with 5 parts sand or fine dried soil. Nearly 900 gram to 1.2 kg good germinated seeds is required for raising of nursery for one hectare plantation. Appropriate time of raising nursery is last week of May or first week of June in north Indian conditions. Sowing should be done in line 7-10 cm apart and care should be taken to avoid deep sowing. It is advisable to cover the seed beds lightly with straw so as to conserve the moisture which may be removed after germination of the seeds. In dry season, it may be necessary to irrigate the nursery beds twice a day by light irrigation. Seedlings are ready for transplantation within 20 to 25 days.

(II) Planting

Transplanting should be done in morning or evening hours, just after the planting, irrigation is required. Cloudy weather and light rain are considered ideal for transplanting. It is recommended to transplant the seedlings in rows and plant spacing of 50 X 40 or 50 X 30 cm, respectively.

Direct sowing/broadcasting

The field should be cleaned of stubbles and weeds, about 10-15 tonnes FYM is required during preparation of the land. In large scale cultivation, farmers cultivate the Indian basil through broadcasting method. In this type, about 2.5 to 3.0 kg seeds are required for 1 hectare of land. One part of the seed and ten parts of the sand should be mixed for proper broadcasting. By broadcasting, spacing of the plants can not be maintained. The thinning is advisable during first weeding for maintaining the population density of the plants.

Irrigation

Irrigation depends upon the moisture of the soil. During monsoon, water requirement of the crop is fully met upto September, thereafter 3 to 4 irrigation are necessary. However, in rainy season, no irrigation is required.

Inter-culture and weeding

The first weeding should be done after 25 to 30 days of the transplantation. The second weeding/hoeing is recommended after one month from first weeding, till plants become bushy to suppress the lower weeds.

In broadcasting type of sowing, weeding is required after 25-30 days from sowing. In first weeding, uprooting is required to unwanted plants to maintain proper population. The second weeding is recommended after one month of first weeding.

Fertilizer/ Manure

Essential oil, the product to be obtained is located in the leaves and inflorescence of Indian basil. Thus, adequate fertilizers are essential for luxuriant growth with good leaves and inflorescence in the crops. In soil of average fertility, about 70 to 80 kg Nitrogen, 40 kg Phosphorus and 40 kg Potash per hectare are advisable. About 25 kg Nitrogen and total quantity of Phosphorus and Potash is applied as a basal dose at the time of plantation. Rest 50 kg Nitrogen is applied in two split doses during the growing season. Due to short duration crop, it is most responsive to chemical fertilizers.

Crop rotation

Due to high productivity and low returns from cereal crops agriculture should be diversified from traditional crops to high value crops like medicinal and aromatic plants. Cultivation of Indian basil as a bonus crop the expenses on cereal crops can be minimized. The rotations are possible both with transplanted and direct planted basil crops. The following crop rotation is recommended:

Basil-Potato – Mint; Basil – Potato; Basil – Mint; Basil- *Brassica* – Mint; Basil- Pea – Mint; Basil-Wheat-Mint.

Harvesting

The basil crop herbage should be harvested when the field is dry and weather is sunny. The crop is harvested at a time of full bloom when lower leaves turn yellowish. The crop may come to full blooming in 2.5 to 3 months after transplanting at subtropical region. In Northern India, only two harvests can be taken. In first harvest, only flowering tops are harvested.

But single harvest is more popular because second harvest of basil takes more time and the proceeding crops will be delayed.

Distillation & Oil Yield

The distillation of the basil is done by Hydro-steam Distillation using field distillation technology. The yield of herbage from a basil crop depends upon a number of factors including the climate, soil properties, varieties used, timely plantation and irrigation, control of weeds, pests and diseases, fertilization, planting and harvesting schedule. Average herb yield of basil is about 20 to 25 tonnes and oil yield is about 110 to 125 kg per hectare. For removing of any moisture, oil should be treated with anhydrous sodium sulphate and decanting.

Uses of Basil oil

The oil has a characteristic warm herbal anise like note with slightly smoky background. Regarding flavour, it is sweet warm spicy anise like with bitter after taste. It is used for imparting green herbal character in modern fragrances, particularly in chypre, floral aldehydic and also in modern fougere types. The general use level in such fragrances is upto 2-3%. It also used in other fragrances for cosmetics as well where herbal warm effect is required. In eau de cologne and green type perfume, this oil gives very interesting effect. This oil also find use in soap fragrances and also in other low cost fragrance as well.

This oil is also a good source for certain isolates and semi synthetic chemicals like linalool and anethol. Estragol (methyl chavicol) and linalool are present in this oil after conversion of estragol in anethol, the anethol and linalool are repented via fractionation. The anethol is used for imparting anise flavour while linalool is used for flavour and fragrance both.

In aromatherapy, it is used as insect repellent and insect bites. Good for bronchitis, cold, cough, earache, sinusitis, etc. The oil is antiseptic, antispasmodic, carminative, digestive, expectorant, prophylactive and insecticide in action. Also, used in fever, flu, anxiety, depression, fatigue and nervous tension.

ECONOMICS OF CULTIVATION OF BASIL

Particulars per hectare	Cost of Cultivation (Rs./ha.)
A. Cost of cultivation and Distillation	
Cost of plant material **(1 kg seeds)**	500
Land preparation for planting	2,000
Fertilizers	1,500
Labour cost	2,000
Irrigation	1,000
Plant protection	500
Weeding (1 time manually)	1,000
Harvesting	1,500
Transportation of herbs	1,000
Distillation charges	2,500
Miscellaneous	500
TOTAL	14,000/-
B. Gross Returns	
Oil Yield (Kg)	110 kg /ha.
Price of Oil (Per kg)	Rs. 350/kg
Total cost of Oil (Per ha.)	38,500/-
C. Net Returns	**24,500/ha./year**

The above cost parameters are based on prevailing market rate of manpower, other materials and produce (essential oil) may deviate in future.

CITRONELLA

(*Cymbopogon winterianus*)

Cultivation Technology of Citronella

The essential oil of citronella is obtained from the distillation of green herb. There are distinctly two types of oil available in the market i.e. Ceylon type and the Java type. The Ceylon type oil is obtained from the species *Cymbopogon nardus* Rendle, (Family-Gramineae), whereas the Java type oil is distilled form the plant *Cymbopogon winterianus* Jowitt. At present, the cultivation of Ceylon type is confined to Ceylon on limited scale. Java type citronella, as it was called mahapengiri in Ceylon was introduced in Java from Ceylon in 1899. Today Citronella Java cultivations is done on large scale in the states of Assam, Meghalaya, Karnataka, Tamil Nadu, Maharashtra, Uttar Pradesh, Bihar and Madhya Pradesh. The present world production of citronella Java oil is estimated to be around 5000 tonnes annually. Indonesia, China, Sri Lanka, Argentina, India, Taiwan and Brazil are the important growers of citronella Java. Citronella Java cultivation in India is spread over to 5000 ha. of land with annual production of about 800-1000 tonnes of oil (Plate 2).

Soil and climate

For optimum growth, deep well drained fertile soil suitable for citronella cultivation, the land should be well drained. Although, pH around 6.0-7.5 is considered to be optimum, successful cultivation of citronella. Although, 150-200 m altitude is considered ideal, the plant has been found grow satisfactorily up to a height of 700 m. Humid climate with regular rainfall offer most favorable condition for plant growth and oil production. Temperatures beyond 38°C and below 20° C do not support good growth of plant. Because of climatic difference, oil yield in north-eastern and southern part of India is higher than those from north India (subtropical climate).

Propagation

Planting material is obtained by dividing the clumps from old plantation. The divided units are called "slips". Depending upon the growth and age of plantation, 20-25 sturdy slips- per clamp can be obtained. Before diggingthe clumps, all the leaves above 20-25 cm. from ground are removed so as to minimize the loss of water though transpiration from the slips. Lengthy roots are trimmed and dead leaves are removed from slips to allow quick establishment and to protect from soil born insect and pests.

Planting

Citronella may be planted in flat beds or ridges, depending upon the texture of the soil. In heavier soils, planting on ridges should be preferred. Slips are placed in holes, made with locally available tools, about 6-8 cm deep and soils around slip is pressed properly. Usually, the planting is done at 60x60 or 60x45cm distance. However, in the areas with large growing period (humid tropics) planting distance can be increased to 70x65 cm or even 80x75 cm. In sub-tropical climate, closer planting 60x45 cm or 60x40 cm is recommended, due to shorter growing period. About 28,000 to 38,000 slips are required for one ha. planting depending upon their spacing. Closer planting helps crop plants to become more competitive against weeds. The best time for planting is the onset of rains. In sub-tropical climates, with assured irrigation, planting can also be done in Feb-March Feb if assured irrigation is available.

Manures and Fertilizers

Being a leafy crop, needs high quantity of soil nutrients. Therefore, it is a general practice to apply sufficient organic manures and fertilizers. In Guatemala and Haiti, it is common practice to apply spent grass (Distilled herbs) to supplement manures and fertilizers requirement, whereas in Indonesia, growers prefer green manuring. The fertilizer requirement of the crop is greatly depends on the fertility status of the soil and the climatic conditions. While in northern plains of India does of 120 kg. N and 40 kg each of P_2O_5 and K_2O per ha is optimum, crop has been found to respond well up 200-250 kg N/ha in poor red laterite soils of Karnataka. The time of planting in first year and before hoeing and after the rainfall in subsequent years, the remaining half of nitrogen should be applied in four equal splits after each harvest.

Irrigation

Citronella requires moist condition for good growth. In sub-tropical areas of north-India where rainy season cover only 3-4 months, the crop requires regular irrigation during rain free period from October-June, and 8-10 irrigations are sufficient to meet the water requirement. In humid part of north-east India where rainfall is well distributed with prolong rainy season, the crop is cultivated in rainfed condition.

Harvesting

The commercial life span of citronella plantation is upto 4-5 years if maintained properly. Depending upon the climatic conditions and management practices, the first harvest from newly established plantation is taken after 5- 6 months from planting. After that, the crop is harvested at 3-4 months interval. Thus, a minimum 4 and maximum 5 harvest/year are taken. Some growers try to take even 6 or more harvests. Harvesting is done with sharp sickle and jerks should be avoided, especially during first harvest, which may cause death of clumps.

Yield and Oil Content

Citronella leaves contain about 0.6- 0.8% essential oil. It varies greatly depending upon crop growth, season of harvest, atmospheric humidity and soil moisture content. Oil content and recovery of oil is lowest during rainy season and the highest during dry months. The production of herb is influenced greatly with the soil, management practices and the climatic conditions. On an average 21-24 tonnes of herb/ha/year is obtained which on distillation yields 150-200 kg. oil. Under very favorable growing condition, it is possible to produce 220-280 kg oil per ha/year. Herb and oil production in the planting year is only about 60%, however, the optimum yield is obtained from second year onwards.

Uses of Citronella Oil

The oil does have sweet, citrus-rosy with terpenic undertone. Regarding taste it is burning-biting and Lemony rosy in dilution. It is used in fragrance of soap, agarbattis, dhoopbatti and detergent. The use level in such fragrances can go upto 15-20%. It is also used nowadays, in fragrances for floor cleaner, glass cleaner etc. Used in flavour for cheaper rose flavour, at times in fruity flavours.

Use a good natural source for Citronella, Geraniol, Citronellol and Elemol. The Citronellol so obtained is utilized for making Hydroxy Citronellol after hydrogenation. The residue left after isolating these isolates is used in cheap fragrance for detergent, soaps and agarbattis.

The oil is also used as a mosquito repellant and sometimes in fragrances for fresherners. Particularly because of its great making power, it is an effective and universally applicable material for technical products, industrial fragrances. With Cedarwood Virginia oil it is used as a popular remedy against mosquito attack in parts before appearances of modern insecticides.

In Aromatherapy, this oil is found to have antiseptic, bactericidal, fungicidal, insecticide etc. Good for oily skin and effective in cold, flu, migraine, headaches and fatigue.

ECONOMICS OF CULTIVATION OF CITRONELLA

Particulars per hectare	Cost of Cultivation 1st year (Rs.)	Cost of Cultivation 2nd year (Rs.)
A. Cost of cultivation and Distillation		
Cost of plant material **(37000 slips)**	18,500	----
Land preparation for planting	2,000	----
Fertilizers	2,000	2,000
Labour cost	2,000	2,000
Irrigation	2,000	5,000
Plant protection	1,000	1,000
Weeding (2 times manually)	2,000	2,000
Harvesting	1,000	2,000
Transportation of herbs	1,000	2,000
Distillation charges	2,500	3,500
Miscellaneous	500	500
TOTAL	34,500	20,000
B. Gross Returns		
Oil Yield (Kg)	150 kg/ha.	150 kg/ha.
Oil Price (Per kg)	Rs. 380/kg	380/kg
Cost of oil produced (per ha.)	57,000/-	60,000/-
C. Net Returns	**22,500/ha./ year**	**37,000/ha./ year**

The above cost parameters are based on prevailing market rate of manpower, other materials and produce (Essential oils) may deviate in future. The returns will be same from 2nd year onwards up to 4th year.

CHAMOMILE

(*Matricaria chamomilla*)

Cultivation Technology of Chamomile

The commercial essential oil of chamomile is known as Blue oil and it is obtained from its shade dried flowers by steam distillation. German chamomile is botanically known as *Matricaria chamomilla* L, apart from this species two other chamomile species found in the world, i.e. Roman or English Egyptian chamomile and Moroccan chamomile. It is used in perfumery as well as medicinal importance like face creams and sun burn and flavoring fine liquor. As medicine, the plant possesses anti-spasmodic, antichloristic and attenuate properties. Extract of chamomile is also useful in stomach disorders, cure of ulcer and healing of wounds. Flowers are extensively used as herbal tea and it is very common in china. The plant is native to Europe and has been used in folk and traditional system of medicine in Europe and Arab for thousand of years (Plate 3).

In India, the plant was introduced by Arabs and has been grown on limited scale around Lucknow for more than three hundred years. The estimated world production of chamomile is about 20,000 tonnes of dried flowers and 7 tonnes of oil. In India, the plant has been cultivated for the production of dried herb, known as "Babuna", and the flowers, known as Gule Babuna in trade. The drug is prescribed by the practitioners of Unani system of medicine in the country. Improved cultivation and distillation technologies of chamomile were Developed by R&D institutions. In the beginning of nineteen, large-scale cultivation of German chamomile was taken up by private growers for the purpose of export of flowers and oils to European countries. In India, cultivation was started in about 40-50 ha. of land with annual production of 30-40 tonnes of flowers. Being a highly labour intensive crop, It is ideally suited for cultivation in eastern Uttar Pradesh and Bihar.

Soils and Climate

It can be cultivated in wide range of soils, except those which are light textured and have low water holding capacity. The cultivation of chamomile has been done successfully on medium and heavy (with high clay content) textured soil, including those of traditional rice growing soils. The plant has also been found to tolerate high degree of sodic soil and hence the soils with pH 8.50 or 9.0 can successfully be utilized for this cultivation. Soils with high water table during winter season do not support good plant grows very well in moderately cool weather. It can be cultivated as winter crop in plains of north India and as summer crop in the hills of J& K, Himachal Pradesh and Uttar Pradesh. The best growing conditions for the plant are 30-32^{0}C temperature and 40-50% atmospheric humidity. Growth and flower bud formation is severely affected when the temperature reaches 38^0 C and above.

Propagation

Chamomile is propagated by seeds for commercial cultivation of crop, in some of the European countries root cuttings are also used. Although, direct sowing of seeds in lines can be done to raise the crop, planting for the commercial production is recommended through raising of seeding in nursery. Raised nursery beds of 4 x 1 or 5 x1 m are prepared by mixing good amount of farmyard manure or compost in the soil. About half to one kg seed, depending upon the germination percentage, is considered sufficient for raising nursery for planting one ha. land. Since, it has been observed that the germination percentage of the seed vary from year to year, it is always better to plant 10-20% more area under nursery than actually required quantity, so that the nursery does not fall short for planting the targeted area. Before sowing, seeds are mixed with 8-10 parts of sand the spread over to the beds uniformly. Seeds are properly mixed in the soil by hand or with some tool. While incorporating the seeds in the nursery, it should be ensured that they are not placed beyond 1 cm. deep in the soil. Sprinkler irrigation is applied immediately after sowing. Thereafter, beds are surface irrigated with slow discharge of water. Sowing of nursery is done in the first 10-15 October in planes and in hilly areas it is recommended in the month of March. In nursery beds, seeds takes 3-5 days for germination and seedlings become ready for planting in the final field after 4-5 week after sowing of nursery. After 3 weeks, nitrogen @ 20 kg/ha is recommended for top dressed in nursery beds for better results.

Planting

After 3-4 weeks, full grown seedlings are uprooted and planted in the final field. The recommended distance of lines are made 25-30 cm. apart and 10-15 cm plant to plant. After planting, the field is flood irrigated with slow discharge of water, so that seedlings are not washed away or uprooted with irrigation water. Delay in the irrigation may cause wilting and death of seedlings in final field.

Manures and Fertilizers

The plant of chamomile has been found responsive to fertilizer application, particularly nitrogen and phosphorus. Nitrogen, phosphorus and potassium in the ratio of 35:30:20: kg/ha. is incorporated in the soil as a basal dose. Depending upon the fertility of the soil, 2-3 top dressings each with 25 kg nitrogen is done. The first top dressing is made after 35-40 days after planting and third after first picking of flowers.

Irrigation

For better plant growth, soil moisture should be maintained at optimum level throughout the growth span. After flowering starts, the crop is irrigated as fortnightly interval to keep the soil moist. Prior to flowering, irrigation at 3-4 weeks interval may be sufficient. Thus, for successful crop production 5-6 irrigation is required.

Picking of Flowers

Picking of flowers is most important and laborious operation in chamomile cultivation. Therefore, the plant is cultivated in areas where sufficient labour force is available at cheaper rate. In sub-tropical climate of north-India, plant starts flowering in the 20th Feb. and continues up to 20th April. During this period 5-6 pickings are done at 10-12 days interval. In most of the chamomile growing countries picking of flowers is done by the family members, especially the women and children. A person can pick 8-12 kg of fresh flowers in a day (8hrs) which on drying may weigh about 2 kg. In large scale plantation the picking is done on contract basis of the quantity of flowers gathered by them. Only the fully developed flowers are picked (buds and fully matured flowers contains less of essential oil). In India, laborers are paid at the rate of Rs 3.00 –Rs 4.00 per kg of fresh flowers picked by them.

Drying of Flowers

The fresh flowers of chamomile contain 70-80% water. The picked flowers are immediately transferred to the drying shed and spread over to ground in very thin layers of 1-2 cm, and allowed to dry till flowers loose 45-65% moisture. At this stage flowers can be dried in thick layers to make available the space for drying of fresh collected flowers. Artificial dryers may also be employed where such facilities exist. In India, the drying is done in thatched huts, having several racks made of bamboo or other locally available material. The flowers are spread on the racks over cloth or thin gunny bag, to allow proper aeration and quick drying. By applying these methods, a farmer can be able to retain the original colour of flowers. In absence of proper drying, the colour of flowers turn to brownish or black, such flowers give very poor oil yield in distillation.

Distillation

Steam distillation method is appropriate for extraction of oil from dried flowers of chamomile, because the flowers require high pressure which can be managed only with the steam distillation. The Cohobation process is applied for improvement of the recovery of oil as certain constituents of oil are soluble in water. Chamomile oil forms crystals and has a tendency to form a deposit and revolves inside the condenser. Therefore, it is advisable to stop the flow of cool water (distillate) into the condenser from time to time, so as to raise the temperature of condenser and to reliquary the oil deposits on the wall of condenser as well as separator.

Oil Content and Yield

The average crop may yield 35-45 quintals fresh flowers/ha within the period of 3.5 months, which on drying gives 9-10 quintals dry flowers. The yield of flower may vary depending upon the fertility of soil, management practices and numbers of picking of flowers. A good managed crop may yield as high as 18-20 quintals dry flowers. The essential oil (blue oil) content in flowers varies from 0.15-1.1%, and largely depends upon the variety grown and the conditions of drying of flowers applied.

Uses of Chamomile oil

The oil has warm herbaceous hay like odor with fresh-fruity undertone. In fresh distilled oil sometimes odor is animalic sweet while in aged oil the odor is pleasant with sweet tobacco warm like.

It is used in fragrance for imparting mellow tone. It is also used in men's fragrances (though in low dosages) particularly colognes. It imparts a warm rich undertone which lasts throughout the stage of evaporation.

It is also used in flavour for mellowing the effect of synthetic ingredients, particularly in tobacco. The use level is upto 1%. It is also used in baked goods and candy flavours.

It do have very vast application in aromatherapy as well. It is analgesic anti-allergic, anti-inflammatory, bactericidal, fungicidal. It is used in skin care for acne, burns, earache, eczema & dermatitis. Also used in hair care. Effective in Arthritis, muscular pain, joint pain & rheumatism. Also used for headache, nervous tension, migraine and stress related conditions.

ECONOMICS OF CULTIVATION OF CHAMOMILE

Particulars per hectare	Cost of Cultivation (Rs./ha.)
A. Cost of cultivation and Distillation	
Cost of plant material (1 kg seeds)	1,000
Land preparation for planting	2,000
Fertilizers	2,000
Labour cost	2,000
Irrigation	2,000
Plant protection	1,000
Weeding (2 times manually)	2,000
Harvesting/flower picking	5,000
Transportation of herbs	1,000
Distillation charges	2,500
Miscellaneous	500
TOTAL	21,000
B. Gross Returns	
Yield of dried flowers (Kg)	900 kg /ha.
Price of dried flowers (Per kg)	Rs. 50/kg
Cost of dried flower (per ha.)	45,000/-
C. Net Returns	**24,000/ha./year**

The above cost parameters are based on prevailing market rate of manpower, other materials and produce (dried flowers) may deviate in future.

GERANIUM

(Pelargonium graveolens)

Cultivation Technology of Geranium

The geranium is called rose geranium and known as first substitute of rose due to rosy smell and chemical constituents present in the oil. The essential oil of Geranium is obtained by steam distillation of matured leaves of the plant. Geranium is botanically known as *Pelargonium graveolens,* which is endemic to Reunion Island. Original habitat of geranium is South Africa where these plants grow in natural condition. Few varieties reached Europe in 1700 but cultivation was started in eighteenth century in Southern France where it was raised as an annual crop due to severe winter conditions. It was introduced in Reunion Islands from France in the late nineteenth century. Almost 50 percent of world production is in Renuion Island. Its sub-species of *P. roseum* and *P. radula* are also used in producing geranium oil in countries like China, Algeria, Morocco and Egypt. Geranium oil is high-grade perfumery oil having rose oil content due to presence of citronellol and geraniol (Plate 4).

In India, a variety of *P. graveolens* used to grow wild in the Nilgiris and Shivery Hills. First plantation of a high yielding variety of geranium imported from Reunion Island was started at Vercaud by a French planter Mensieur Sens in the early 20th century. Two varieties of Geranium *i. e.,* Algerian or Tunisian and Bourbon of Reunion were being cultivated in Yercaud (Shivery Hills) and Nilgiris. Former type is grown in the Nilgiris and is slender with dark pink flowers and more sturdy and it is unsuitable in wet conditions while later having light pink flowers is more suitable in wet conditions.

Soil and climate

Geranium grows well in temperate, tropical and sub-tropical climates at various altitudes. However, it thrives best in sub tropical climate. It is sensitive to winter rains, susceptible to water logging. It comes up best in porous, open and slightly calcareous soils having pH range of 6.5 to 7.5. Evenly distributed rainfall of 1000 cm to 1500 cm is considered essential. Heavy rainfall condition accompanied bymisty weather brings disease resulting in the loss of foliage. The plants are susceptible to frost. Maximum yield is obtained when temperature range between 15°-18°C throughout the growing period.

North-Indian region which enjoys cold winters and assured irrigation facilities offer luxuriant growth from November to April and gives yield comparable to hilly areas of South. At present India produces 05 tonnes of oil from 950 ha in Nilgiris and 45 ha in Palani Hills, Bangalore and Hyderabad while the national requirement is 150 tonnes which is met through imports. Therefore, it is imperative to identify new areas suitable for cultivation of Geranium. In north Indian plains the resource plants from pots are transferred to the field in October. Cuttings are prepared from these resource plants from mid November onwards at 10-15 days interval. These cuttings are directly planted into fields at spacing of 50 x 50 cm. The fields are kept moist with light irrigations. Nursery could be set up in Poly Packs as well.

Propagation

Planting material is prepared from cuttings obtained by cutting green portion of the stem having 3-4 nodes. Green cuttings put in moist soil root easily to the extent of 70-80%. Brown stem cuttings have high mortality. Application of seradix B-1 (1 BA@ 50-100 ppm for 5 minutes) enhances rooting of brown stems. In North India, cuttings are planted in the field during late November to 1st week of December followed by 2-3 light irrigations. In South India, planting material is prepared in nursery and subsequently transplanted in thefield during May-July. Spacing of 60x60cm or 60x50 cm is followed for a good growth and 27-28,000 cuttings are required per hectare of land. During summer 2-3 irrigation are required in South while in North 7-8 irrigation are required. Irrigation twice a week is preferred in initial stage.

Manure and fertilizers

About 25-30 tons of FYM is applied per hectare at the time of preparation of land P:K: @ 80:60 are also applied as basal dose.

Subsequently, 100-120 kg/ha of Nitrogen is applied as top dressing in 4 split doses during growth of crop. Two or three 0.5% sprays of micronutrients during vegetative growth at intervals of 15 days would give good growth of foliage.

Weeding

The crop may require one weeding after about 30 days from planting and one hoeing after each harvest.

Disease and Pests

Insect, pests except termites do not easily affect geranium but regular spray of fungicides and bactericides is required to avoid possible attack of pathogens. Sometime geranium plants are attacked by termites, which could be prevented by spray of chloropyriphos 600-800 ml/ha diluted in water.

Harvesting

In the year of planting 1 or 2 harvests are taken but from second year onwards the leafy shoots are harvested 3-4 times in a year. The tender leaves are harvested at the beginning of the flowering season when the lemon like odour of the leaves change to faint rosy odour and appearance of pinkish flowers. Harvesting is done in the early morning after sunrise with a sharp sickle. Only green portion of the plant is harvested leaving smaller branches in tact, which helps in regeneration, and second crop is obtained after 45-50 days.

Yield

Good soil and effective care can give yield of 25-27 tons of fresh herbage. Second yield gives 20-25 tons of foliage. Distillation of fresh leaves or slight withering helps to improve the yield. Normally 0.12-0.16 percent oil is obtained. Farmer can ascertain the oil content by distillation of small quantity of herbage in "Clevenger" apparatus. A farmer can get 40-50 kg oil per ha./year.

Uses of Geranium Oil

It has green, leafy-rosy, minty odour with lasting sweet rosy note. It is used for imparting green rosy note in fragrances for personal and cosmetic products. The terpenless oil is used in high class fine perfumery.

It is also used in gutka, tobacco and pan masala and also for as rose flavour for different food products.

The oil is good for acne and oily skin and for stress related conditions. The deoiled material can be powdered and can be used for agarbatti, dhoopbatti and face packs.

Isolates- The rhodinol isolated out of geranium oil is well known in fragrance & flavour industry and is a highly priced natural aroma chemical. Rhodinol is also used extensively in perfumery as a base for rose, muguet, carnation, apple blossom, and many other perfume types. Its floral sweetness, delightful freshness and delicate warm rosy tonalities blend well with the odors of hydroxy citronellol, linalool, geraniol, dimethylbenzyl carbinol, Cinnamic alcohol, cyclamen aldehyde, phenyl ethyl alcohol, geranyl esters, linalyl esters, etc.

ECONOMICS OF CULTIVATION OF GERANIUM

Particulars per hectare	**Cost of Cultivation 1st year (Rs.)**	**Cost of Cultivation 2nd year (Rs.)**
A. Cost of cultivation and Distillation		
Cost of plant material	13,500	----
Land preparation for planting of seeds	2,000	----
Fertilizers	2,000	1,000
Labour cost	2,000	2,000
Irrigation	5,000	6,000
Plant protection	1,000	1,000
Weeding (2 times manually)	2,000	2,000
Harvesting	2,000	2,000
Transportation of herbs	2,000	2,000
Distillation charges	8,000	10,000
Miscellaneous	500	500
TOTAL	40,000	26,500
B. Gross Returns		
Oil Yield (Kg)	40 kg/ha.	45 kg/ha.
Oil Price (Per kg)	Rs. 3,000/kg	3,000/kg
Cost of oil produced (per ha.)	1,20,000/-	1,35,000/-
C. Net Returns	80,000/ha./ year	1,08,500/ha./ year

The above cost parameters are based on prevailing market rate of manpower, other materials and produce (Essential oils) may deviate in future.

JASMINE

(*Jasminum sambac*)

Cultivation Technology of Jasmine

There are two types of Jasmine found in India, one is *Bela* (*Jasminum sambac*) and other is Chameli (*Jasminum grandiflorum*) are an aromatic crops having an important place in essential oil industry. It belongs to family *Oleaceae.* In the world, about 2000 species known, only *Jasminum grandiflorum* and *J. sambac* are preferably used in perfumery industry. *Jasminum sambac* is an erect shrub with 5-7 dark green leaflets. Flowers are born auxiliary or terminally in cymes (Plate 5).

Soil and climate

Jasmine grows well in good-drained soils. Sandy-loam to clay-loam soils is best suited for the growth of Jasmine. Water-logged soils are not suitable for the growth of the crop and may prove detrimental for plant growth. It prefers mild climate that of Northern plains and Southern part of India.

Propagation

Jasmine is propagated by cuttings during March. The cuttings should be atleast 20-25 cm. long. Each cutting should have three inter-nodes. Middle inter- node should be placed in 1 cm deep in soil. The Nursery is prepared in raised nursery beds to avoid poor establishment in main field after transplanting. The most suitable time for planting stem cuttings for raising nursery is July to August and January-February, 8-10 months are required for growing cuttings in nursery beds before transplanting.

Transplanting

Transplants should be done at the spacing of 1.5 - 2 m x 1.5 - 2 m. About 3000-4000 plants/hectare can be planted depending on spacing of the plant. The jerks and pulls should be avoided while transporting the planting material. The sapling should be planted along with earth ball after removing the polythene cover. The surrounding soil then pressed and irrigated. Planting is generally done in monsoon, winter and in mild summer. Hot summer and severe winter should be avoided for transplanting.

Irrigation

Irrigation is not required in case of crop planted during monsoon season. In case of crop planted during winter or mild summer, irrigation should be given at an interval of 5 to 7 days till the saplings are properly established (about 4 months). After this period, plant may be irrigated once a fortnight during summer and once a month during winter. One year after transplanting, jasmine develops tolerance to lack of moisture. Established plants of jasmine could survive well absolutely without irrigation. But, water stagnation conditions lead to the death of the plants. So a proper drainage system is required for good plantation.

Manure and Fertilizers

In Jasmine fertigation is done by plant to plant basis, it needs 10-15kg. FYM with 70-80g.N, 120-150g P_2O_5 & 150-200g K_2O per year applied in 3 split doses, (January, April & August).

Pests, diseases and their control

Jasmine is reported to be affected by a number of diseases and pests. Leaves are often susceptible to leaf spot disease resulting in defoliation. Bud-worms eat away the tender shoots of the plants. Spraying of parathion helps control this disease. Spraying of trithion controls the red mite.

Pruning

Pruning is done from 2nd year onwards of the planting. A Jasmine plant has commercial life span of about 10-15 years. Pruning should be done to a height of about 60-70 cm from the ground level. Jasmine should be pruned in December- January. It helps the shoots in retaining maximum flower yield.

Weed control

Weeds compete for nutrition and moisture and cause hindrance in harvesting and attract pests and diseases. 3-4 times hoeing in a year is beneficial and checks weed growth. Around January or February every year, the soil should be ploughed well.

Harvesting/picking of flowers

Jasmine starts flowering in the second year from planting. The flowering period ranges between April and May and from August to November. Harvesting/plucking of flower is done preferably in the morning as flowers contain maximum aroma at that time. About 25 to 30 persons/ha/day are required for plucking the flowers. Warm weather and sunshine result in better quality flowers that are heavily scented. Cool and rainy weather adversely affects the yield of flowers.

Extraction/Distillation

The solvent extraction method is the commonly used for the production of Jasmine concrete. In this process, non-polar organic solvent preferably n-hexane is used. Flowers are dipped in a solvent extraction unit together with n-hexane. The hexane enriched with volatile fragrant compounds is led to vacuum distillation for removal of solvent. Jasmine fragrant compounds as well as undesired plant waxes get extracted in this procedure. The waxes present in the concrete (about 50%) makes this product for limited use in perfumery Industry and hence undesirable in the form concrete. So the concrete is treated 4-5 times with high grade absolute alcohol and refluxed at 54°C. The alcoholic solution is cooled from -25 to -30 to precipitate the undesired soluble waxes which are filtered off. The wax free solution is concentrated by distilling out the alcohol at 35°C at 20mm/Hg pressure. The end product is semi-viscous brown coloured liquid known as absolute or otto of Jasmine. The total yield of concrete is about 10-12 kg/ha./year.

Uses of Jasmine

Similar to tuberose, absolute of Jasmine is also made. Attar of Jasmine sambae (Bela) is made not of Jasmine grandiflora (Chameli). The odor is warm, floral, sweet, fatty floral with slightly herbaceous with fruity undertone.

The odor and nuance in undertone varies according to origin, production, age and method of purification from concrete. Aged material gives fatty note as well.

Regarding used one can cite the old saying 'No perfume without Jasmine.' The Jasmine-rose complex forms the more important part of numerous 'grand perfumes' as well as quite ordinary perfumes and bases. Nearly in 80% cases of floral note is imparted by jasmine, though in small amount. Hence, looking into cost/impact ratio jasmine absolute doesnot appear to be an expensive fragrant raw material. Jasmine concrete is used in soap fragrance where its effect can be perceived in low dosages. Hence, it is used sometimes in medium-to premium soap fragrances. When used properly absolutes helps in 'rounding off' chemical notes giving naturalness and a warm floral undertone. In chypre bases this is used a lot.

Also used in flavours due to its warm floral taste. Occasionally used in liquors and pastries. In baked goods & ice creams flavour it is utilized as well.

In aromatherapy its action is mild analgesic, anti-inflammatory, antiseptic, antispasmodic and sedative. It is used for dry, sensitive and irritated skin. Good for muscular spasm and sprains. Useful in depression, nervous exhaustion and stress. Its action is stimulant and similar to that of caffeine.

Particulars per hectare	**Cost of Cultivation 1st year (Rs.)**	**Cost of Cultivation 2nd year (Rs.)**
A. Cost of Cultivation and Distillation		
Cost of seeds/plant material **(3000 plants)**	15,000	----
Land preparation for planting of seeds	4,000	----
Fertilizers	2,500	2,500
Labour cost	4,000	5,000
Irrigation (6)	4,000	5,000
Plant protection	500	500
Weeding (2 times manually)	2,000	2,000
Harvesting/flower picking	5,000	8,000
Transportation of flower	2,000	4,000
Distillation charges	15,000	20,000
Miscellaneous	1,000	500
TOTAL	55,000	47,500

B. Gross Returns		
Concrete Yield (Kg)	7 kg/ha.	10 kg/ha.
Price of concrete (Per kg)*	Rs. 12,000/kg	12,000/kg
Cost of oil produced (per ha.)	84,000/-	1,20,000/-
C. Net Returns	29,000/ha./year	72,500/ha./year

ECONOMICS OF CULTIVATION OF JASMINE

The above cost parameters are based on prevailing market rate of manpower, other materials and produce (Concrete) may deviate in future.

LAVENDER

(*Lavandula officinalis*)

Cultivation Technology of Lavender

Lavender oil is used in high class perfumery and soaps due to its refreshing and sweet odour and it is blended with many other essential oils. The major producers of lavender oil are France, Italy, Russia, Hungary, U.K., Bulgaria, Austria, China, U.S.A and few parts of India especially in hilly region. India is importing more than 50 tonnes of lavender oil for use in cosmetic and soap industry. Bulgaria produces 100-120T, France 50-55 T and Russia 35-40 T of Lavender oil. During pre-independence era Britishers tried to cultivate the plant in Kashmir and *Ooty* but they did not make any attempt to commercialize the technology. The Drug Research Laboratory, J and K tried to introduce the plant in various parts of Kashmir valley and results were encouraging as the oil produced was of international standards but again it was cultivated on small scale as a non-commercial activity. The plant grows well in sub-tropical areas and flourishes well in cold/hilly areas where summer as also cool-winter (0-8^{0}C and summer 25-30°C) prevails. Its roots go deep into the soil. It grows well in sloppy lands and checks erosion to a great extent. The best crop and oil yield are obtained in neutral and alkaline soils (pH 7.0- 8.2). It is resistant to frost and drought. It prefers open sunny soils for proper growth. The plant could be cultivated on wasteland/eroded soils. Keeping these facts in view, it should be attempted in Himachal Pradesh, Uttarakhand, hills of south India. (Plate 6)

The genus *Lavendula* belongs to family Labiatae comprises about 28 species excluding hybrids of perennial aromatic herbs or shrubs. The other varieties generally cultivated are:

1) ***Lavandula augustifolia*** Mill sub sp. *augusti folia* (syn. *Lavandula officinalis* chaix)-shrub 50-80 cm height
2) Spike lavender, L - *latifolia* Mill
3) Lavendin – *L. intermedia* -it is hybrid between true Lavender and spike Lavender.

Lavender is cultivated through seeds as well as rooted cuttings. The propagation through seeds is economical and easier but there is danger of cross-pollination which results in variability of quality of oil. The vegetative propagation ensures a uniform quality of oil though the process is slower, labour intensive and expensive.

Seeds are sown in beds of 1m width in November-December either through broadcasting, or in rows 10-12 cm apart for raising nursery. Seeds are mixed with the sand or rotten compost, wash before sowing and 10-12 kg seeds are required for 1 ha. area. The sowing depth is 1-2 cm. The seed hole is covered with FYM or fine sand and left as such during winter. The seeds germinate in spring (April) when temperature is rise. The nursery beds are lightly sprinkled and kept moist. The seedlings should be pruned periodically to avoid follicle formation.

Lavender is also propagated by rooted cuttings inserted into sandy soil in open, protection being given in cold weather. Light watering is given to establish the plants but during day weather Netlon may be used to protect the plants.

Transplantation can be done either in autumn or spring. Plant to plant distance is 20-30 cm and rows 40-50 cm apart. About 65-66,000 plants per ha. give the highest yield. Earthing of the soil around the seedlings give good results.

Fertilizers and manures

A basal dose of 20 kg Nitrogen and 20 kg Phosphorus and 20 kg of potash with 10-15 t. FYM per ha. is applied before planting. Two split doses of nitrogen is required for good yield in first year. Subsequently, three split doses of Nitrogen @ 15 kg each during growing season in May, June-July are applied.

Irrigation

For better plant growth and flower production lavender requires frequent irrigations. However, the regular rains are sufficient for the production of lavender.

Weeding

Regular weeding and hoeing is done to de-weed the field and to loosen the soil for proper growth.

Regeneration

The commercial life span of Lavender plant is 15-20 years. The yield of oil is optimum after 2nd year onwards and reduced after 15 years which necessitates re-generation during restive period of plant i.e. November-February. The plants are cut 4-5 cm above the ground and heavy dose of fertilizers is provided. Regeneration restores production capacity of old plantation which otherwise should have been uprooted and re-planted.

Harvesting

Plant starts flowering earlier in winter and low altitude areas and later at higher altitude. The highest yield is found in November to December. Flowers are cut off with stem lengths of less than 10 or 12 cm because of absence of oil in stems and leaves. Production of higher quality lavender oil depends on a number of factors such as degree of blossoming, picking during appropriate hours, atmospheric conditions, methods of harvesting, storage and flower transportation. Harvesting should start at 50% blossoming and continued upto 75-90% in florescence and upto the stage when flowers end blossoming. Harvesting should be done on warm sunny days. As such if rain or fog appears, harvesting should be discontinued and resumed on sunny days.

Distillation & Oil yield

The lavender oil is obtained by steam distillation only. The distillation process takes 3-4 hours from fresh flowers. Flowers should not be stored for long time, which results in low yield of oil. Stainless Steel distillation chamber, vapour line and condenser should be used. The oil content is 1.2 -1.5% and a good crop can give oil yield upto 40-50 kg/ha.

Uses of Lavender oil

The odour is sweet, refreshing floral-herbaceous-camphoraceous having balsmic background. The lavender oil is a well known oil in perfumery industry. It is used in high class fine fragrances and for fragrances in cosmetic products. This oil is also used as fragrances in room fresheners. The oil is used in making colognes, lavender water, fougere, chypres, floral and semi floral perfumes. The oil is used when we need diffusion in such kind of fragrances, however, lavender absolute finds a place when we need stability in such fragrances. The oil and absolute both are good for use in soaps, perfumery as well.

This is the oil which is responsible for birth of Aromatherapy, hence, found use in skin care (both dry and oily skin) and burns. Also good for anxiety, muscular aches and pains. In low doses, it acts as a very good sedative while in high dose acts as a stimulant. This is most wildly used oil in aromatherapy and can be used even on child skin. This is the oil which can be used even without dilution with a carrier oil. The oil is antiseptic and pain reliever. Good for sun burn and sun stroke, effective in asthma and bronchitis. Helpful in treatment of head lies. Very effective for tension headache, relaxation and anxiety. It is also find its use as a mosquito repellant.

ECONOMICS OF CULTIVATION OF LAVENDER

Particulars per hectare	Cost of Cultivation 1st year (Rs.)	Cost of Cultivation 2nd year (Rs.)
A. Cost of cultivation and Distillation		
Cost of seeds/plant material (12kg/ha.)	12,000	----
Land preparation for planting of seeds	3,000	----
Fertilizers	2,000	1,000
Labour cost	5,000	5,000
Irrigation (2)	3,000	3,000
Plant protection	1,000	1,000
Weeding (2 times manually)	2,000	2,000
Harvesting	2,000	2,000
Transportation of herbs	2,000	2,000
Distillation charges	7,500	7,500
Miscellaneous	500	500

TOTAL	40,000	24,000
B. Gross Returns Oil Yield (Kg) Oil Price (Per kg) Cost of oil produced (per ha.)	 40 kg/ha. Rs. 2,500/kg 1,00,000/-	 50 kg/ha. 2,500/kg 1,25,000/-
C. Net Returns	60,000/ha./y ear	85,000/ha./y ear

The above cost parameters are based on prevailing market rate of manpower, other materials and produce (Essential oils) may deviate in future.

LEMONGRASS

(*Cymbopogon flexuosus*)

Cultivation Technology of Lemongrass

Lemongrass (*Cymbopogon flexuosus*) belongs to family Gramineae, grows wild in many tropical and sub-tropical parts of Asia, Africa and America. This is indigenous to India and large variability occurs in the Western Ghats comprising Kerala, Karnataka and Tamil Nadu besides Sikkim and Arunachal Pradesh. Lemongrass is a perennial aromatic grass, which yields essential oil having lemony odour due to high citral content (75-85%) in its oil. Three types of essential oils are popular in trade i.e. East Indian, West Indian and Jammu lemongrass. It is about 2-2.5 meters tall with dark green or yellowish green leaves. The oil is present in oil glands presented in leaves tissues. Leaf sheath are glabrous, hairy at the junction blade, leaf blade is one meter long and 1.5 cm. wide (Plate 7).

Soil and Climate

Lemongrass is a very hardy crop, which can be adapted to a varied nature of soil and climate. It grows best in well-drained sandy loam and medium loam soils having good water holding capacity. It can be grown in a varied pH (6.5 to 8.5) range of soils. But, it prefers neutral pH soils, which supported good growth of plants. It can also be grown in soil with poor fertility, marginal and sub-marginal land. It grows best in warm and humid climate with bright sunshine and well distribution of rainfall throughout the year.

Propagation

Lemongrass is propagated through vegetative parts called as slips. A slip is separated from a bunch of tillers of one year old plant population. Propagation is done in the month of July / August during monsoon season. However, it can also be planted in February-March if farmer have assured irrigation system. One or two slips are planted in a hole at a distance of 45cm. and 60 cm. apart. By this spacing about 37,000 slips are required for one ha. of land.

Fertilizers requirement

It is a leafy crop and responds well to use of fertilizers and manners in the soil. Fertilizers applicationin lemongrass varies from soil to soil and region to region. The fertilizers and manure requirement in heavy soils are less than the soils of poor fertility. In an average fertile soil, 120 kg. nitrogen, 60kg. phosphorus and 40 kg. potash is recommended; 60 kg. of nitrogen, all phosphorus and potash is applied at the time of planting and rest amount of nitrogen is applied into three equal splits during growth period of crop.

Weed Management

Weed infestation is avoided at the initial stage of crop, it can be done by manual weeding or by using chemical herbicides. Early stage of growth weed competition is more than the developed crop. Weed control can also be done by using organic mulching, which can be done by using waste of citronella and lemongrass itself. Mulch is spread out between the lines of crop, which prohibit the growth of undesired weeds.

Irrigation

It is an important factor, which is responsible for healthy plant span. If crop is planted in July/August (Monsoon season), no need of irrigation in the crop. Generally, irrigation is required during dry month of the year i.e. March to June. For successful growth, 6-8 irrigation's are required in a year.

Disease and Pests

Lemongrass is much hardy and resistant to disease and pests problem, however, a number of diseases are reported in which leaf blight is more harmful to the crop. It is caused by a fungus known as *Collatotricum graminicola*. At previous stage of infection, small brown patches appear on the leaves surface and these elongated through the age. The disease can be controlled by using dithiocarbamate.

Harvesting

Harvesting is done in sunny days by a sharp sickle, jerks and pulls are avoided for better generation of earlier crop. First harvest of lemongrass is obtained in 120 days of planting and subsequent harvest is obtained after 60-80 days of previous harvest. In case of old cultivars, it is not applicable; it can be harvested within 80-120 days under good management condition in improved variety. In a year 4-6 harvests can be obtained from good crop.

Distillation

Distillation can be done by steam or hydro-steam distillation method. The later is more common among the growers because it is too cheap than the earlier. The distillation unit is run by using waste of lemongrass as a fuel so as to made distillation process cheaper. The distillation is over with in 4-6 hours and 0.6- 1.0% oil yield is recovered in normal condition. A farmer can get 150 to 200 kg oil from one ha. land in one year by cultivating improved varieties.

Uses of Lemongrass Oil

The oil has a strong fresh grassy, citrusy lemon peel like odour. While in taste, it is bitter-burning and gives pleasant lemon like taste in dilution.

The oil is used is fragrances for soaps and detergents in low dosages being unstable in alkaline media. The oil is a major ingredient in industrial household fragrances. In flavours, it is used in hot beverages (pain balms), also in baked goods, chewing gums and pudding.

The lemongrass oil is good source for natural citral, it can be further used in soft drinks, flavours, cologne fragrances. The citral can be converted in variety of pseudo-ionones from which alpha-ionone, beta-ionone methyl- ionone, methyl ionone gamma, etc. can be made, which are very extensively used in fragrances and flavours. This citral is also a sources material for vitamin-A synthesis.

From aromatherapy point of view, it is analgesic, antidepressant, antimicrobial, antiseptic, fungicidal, insecticide, sedative, astringent in action. This can be used for acne, scabies, muscular pain, insect repellent, stress related conditions, headache, etc.

ECONOMICS OF CULTIVATION OF LEMONGRASS

Particulars per hectare	**Cost of Cultivation 1st year (Rs.)**	**Cost of Cultivation 2nd year (Rs.)**
A. Cost of cultivation and Distillation		
Cost of plant material **(37000 slips)**	18,500	----
Land preparation for planting	2,000	----
Fertilizers	2,000	2,000
Labour cost	2,000	2,000
Irrigation	2,000	5,000
Plant protection	1,000	1,000
Weeding (2 times manually)	2,000	2,000
Harvesting	1,000	2,000
Transportation of herbs	1,000	2,000
Distillation charges	2,500	3,500
Miscellaneous	500	500
TOTAL	34,500	20,000
B. Gross Returns		
Oil Yield (Kg)	120 kg/ha.	150 kg/ha.
Oil Price (Per kg)	Rs. 400/kg	400/kg
Cost of oil produced (per ha.)	48,000/-	60,000/-
C. Net Returns	**13,500/ha./ year**	**40,000/ha./ year**

The above cost parameters are based on prevailing market rate of manpower, other materials and produce (Essential oils) may deviate in future. The returns will be same from 2nd year onwards up to 4th year.

MENTHOL MINT

(*Mentha arvensis*)

Cultivation Technology of Menthol Mint

Mentha arvensis L. is an aromatic plant belongs to family Labiatae/laminaceae, commonly called as corn mint, Japanese mint and menthol mint. It is a perennial sub-shrub growing upto height of 70-100 cms. It has many branches of green or light purple in colour. Leaves are opposite, lanceolate and upto 10cm-length and 3-4 cm. wide when fully developed. The plant is developed through vegetative means called as stolons/ suckers. It is an underground part of plant. (Plate 8)

Soil and Climate

Basically, it is a temperate plant but, it has a wide climatic tolerance for its growth. It can be grown in warm temperate, sub tropical and semitropical conditions of climate. For successful growth it requires deep, fertile loam or clay loam soils with 6.5-7.5 pH range. Soil with organic matter and good water holding capacity considered ideal, the day/night temperature combination should be 30° C/18° C for its successful growth.

Propagation

Usually, propagation of Japanese mint is done by underground parts of roots called as suckers/stolons. Two methods for planting are prevalent among growers i.e. nursery raised/ transplanted method and direct sowing of suckers in the main field, nursery raised method is much more common. About 90 kg suckers are required for nursery plantation for one ha. of land however, 400 kg suckers are required for direct plantation for one ha. In mint growing areas, about 65% of mint growers are using this method. The best time of planting of mint is 25th Jan. to 15st March.

Planting & After care

Planting of mint crop is done by two above methods. After planting the crop weeding, irrigation, disease control are main components which are responsible for healthy plant growth. Weeding can be done manually or by spraying of herbicide. For satisfactory growth of crop irrigation schedule should be at proper stage of crop growth. Generally, 12-14 irrigation are required for proper development and maturity of the plant, it is also depend on the climate and soil structure.

Manure & Fertilizer

Menthol mint responds well to nitrogen fertilizer application, in average conditions, application of 150 kg. nitrogen, 60 kg. phosphorus and 40kg. potash per hectare per year is recommended. One third of nitrogen and all phosphorus and potash are applied before planting as a basal dose and rest of nitrogen is applied in 2/3 split does after each harvest.

Cropping pattern

Different cropping systems are being used by growers in different regions of India, depending upon climatic variation. Usually, paddy- potato-mint, paddy-mint, paddy-mustard-mint and Basil-potato-mint, paddy-wheat-mint, Basil-chamomile-mint cropping pattern are most common among growers.

Pests & Disease and their control

It is not a severe problem in mint growing areas but in some areas downy mildew (*Erysiphe cichoracearum*), stolon rot (*Rizoctonia bataticola*) and leaf spot (*Curvularia lunata*) infection has been reported. The downy mildew diseases can be controlled by spray of 0.3% wettable sulphur or 0.05% karatane. However, leaf spot can be controlled by spraying of dithane Z-78 (0.3%) in early stages. These problems are caused by fungus pathogen. The symptoms of disease appears in the month of April and May in the form of yellowish brown blisters on the upper surface of the leaves, which are covered with a powdery substances. Disease free plant/stocks are advised to be used for next year plantation.

Harvesting and yield

The crop becomes ready within 100-110 days for first harvest and 60-70 days for subsequent second harvest using late sowing and early maturing varieties. Two harvests are being taken by majority of the growers in mint growing regions. The highest oil content has been reported to be at 0.8 to 1.2% of full blooming stage. Harvesting is done mechanically or manually by cutting the herbage 8-10 cms. from ground level. The average oil yield is about 150-200 kg. obtained from one ha of land form two harvests.

Uses of Menthol Mint oil

The oil does have grassy herbaceous minty note with resinous woody base. While it taste like bitter minty with cool mouth feel. The oil is mainly used as natural source for l-menthol by crystallizing the oil in deepfreezer. The de-mentholized oil (DMO) so produced as a by product of this crystallization process is used as such into flavour and also as source for liquid menthol, cis-3 hexenol (leaf alcohol) which are very important in fragrance and flavour formulation. A number of other isolates like iso-menthone, neo-menthone are also obtained from DMO which are very useful in traces in fragrance formulations. The oil due to its rough some what slightly bitter green or sharp, less balsamic herbaceous character than peppermint oil cannot found it's use as such. It needs to be blended properly or otherwise used as source for crystal menthol and DMO.

The isolates so obtained out of mint oil or after proper blending does find its use in flavouring candy, chewing gums , chocolates filling, toothpaste, mouth washer, gargles, mouth fresheners. It is very intensively used in flavoring pan masala, tobacco, supari as well. Last but not the least pharmaceutical preparation, hair oil, mentholated cigarettes are also into the list of its uses.

It is also used as a cheaper substitute to peppermint oil in number of flavours and due to its cooling and paralyzing effect in taste, it is difficult to identify it.

ECONOMICS OF CULTIVATION MENTHOL MINT

Particulars per hectare	**Rupees**
Cost of cultivation and distillation	
Cost of suckers (400kghs/ha@ Rs20/kg)	8,000
Land preparation for planting of the seeds	3,000
Farm yard manure and NPK	2,500
Labour and tractor cost	2,000
Irrigation (12 times)	2,500
Plant protection	500
Weeding (3 times manually)	1,500
Harvesting (2 times)	2,000
Transportation of distillation unit	1,000
Distillation charges @ Rs 200 /batch)	3,500
Miscellaneous	1,000
Total	**27,500/-**

A. Gross returns	
Oil yield (kg)	150 kg/ha
Oil price (per kg)*	Rs 450/kg
Cost of oil produced (per ha)	Rs 67,500/-
B. Net returns	**Rs 67,500-27,500 = Rs 40,000/-**

*= The selling price of the menthol mint oil varies from season to season and year to year based on the international supply and demand situation. In 2008-09, menthol mint oil price fluctuated between Rs. 500-850 / kg.

MARIGOLD

(*Tagetes Spp.)*

Cultivation Technology of Marigold

Marigold belongs to family Asteraceae and commonly known as *Genda* in Hindi. There are three species which are commercially cultivated in India.

a. Tagetes minuta

b. Tagetes petula

c. Tagetes errecta

The marigold flowers have been used since time immemorial in religious and social functions for decoration, garlands, extraction of essential oil and preparation of herbal edible dyes. It is a well-known companion crop with vegetables and ornamentals. Marigold gained popularity among growers due to easy adoptability and survival. The plant is reported to be of medicinal value and finds use in Indian System of Medicine. The essential oil present in different species is being used in perfume and Attar industry (Plate 9).

All types of marigold petals contain essential oil which is extracted by steam/ hydro-steam distillation. The oil is also extracted from its leaves and whole plants. The commercial essential oil is produced from both wild populations and cultivated plants grown in several temperate and semi-temperate parts of Asia, Africa, Europe, America and Australia. Some marigold cultivation is carried out in lower reaches of Himalayas in the states of Jammu and Kashmir, Himachal Pradesh and Uttar Pradesh and Madhya Pradesh. The demand of essential oils of Tagetes spp. is high in demand for preparation of natural insecticide, perfumery, in pan masala industry and for preparation of natural colourants, considering its wide usage it should be propagated widely in plains of north India. Farmers are mainly cultivated African marigold (*T. minuta*) for its commercial essential oils. In this plant, there is no real flower form, only spikes are presented and that is the indicator for harvesting. Whole plant is used for extraction of essential oil using distillation technology. The detail cultivation technology of *T. minuta* is as under:

Soil and climate

Tagetes minuta can be cultivated on a wide range of soils from sandy loam to clayey loam soil. Well-drained deep loam soil rich in organic matter having pH 4.5 to 7.5 is ideal for its cultivation. Temperate as well as semi-temperate climate is suitable for its cultivation. African marigold requires cool climate for seed germination and prolonged summer for optimum vegetative growth and flowering. The temperate region is best for its good chemical constituents.

Propagation

In sub-tropical areas, the crop can be directly planted through seeds during September-October, while in temperate areas the nursery is raised in March-April for later transplantation. The well rotten Farm Yard Manure is mixed with top layer of soil. The seeds are sown in the soil 1-1.5 inches deep surface and then covered with fine soil or farm yard manure and light irrigation is applied. Regular light irrigation is given from time to time as and when required. After 25-28 days from nursery sowing, seedlings are up-rooted from nursery bed and transplanted at the spacing of 45-30 cm. Surface light irrigation is given just after transplanting of saplings in final field. About 1.5-3.0 kg of seeds is required for one hectare of land by direct sowing and 800gm. - 1.0 kg of seeds are required for raising nursery. Undesirable or weak plants can be uprooted during first weeding after 30-40 days of seed sowing.

Irrigation

The crop is wild in nature so its does not require frequent irrigation but 7 to 8 irrigations are required for good growth of crop in north Indian plains however, 2-4 irrigations in hilly areas are sufficient.

Manures and Fertilizer

Due to leafy crop, the application of fertilizers responds well in this crop. During preparation of field, 12-15 tons of well rotten FYM should be applied before final ploughing. Apart from FYM, Nitrogen 80 kg, Phosphorous 50 kg and Potash 35kg is required for good growth and yield. Nitrogen is applied in three split doses, first dose after 30 days and second after 30 month of first dose and final dose at the time of flowering, while P and K are used as basal dose.

Harvesting and Yield

Harvesting period is decided when the crop is in full bloom and has started towards maturity. The crop grown in September-October in plains is harvested in March-April while in hills crop planted in March-April is harvested in September-October. The commercial life span of this crop is six-month. The harvesting is done 15-20 cm above from the ground level. The cultivation of improved varieties, in north Indian plains gives biomass of 25-30 tonnes/ha which gives 50-60 kg essential oil. The herbage should be distilled between 2-3 hours from harvesting.

Uses of Marigold Oil

Odor is herbaceous, sharp floral, green fruity. If oil is older it will lack fruity top note. Care has to be taken to store this oil, otherwise it oxidizes and resinifies if exposed to air, moisture and heat.

The oil is used in small dosage due to its strong aroma. It is used in fragrances particularly herbaceous, Lavender, fougere and chypre. Also used in floral for imparting green-herbaceous note giving naturalness. Such florals are hyacinth, violet, jasmine and gardenia. Gives interesting effect in modern aldehydic bases. It blends well with naturals like Sandalwood, Bergamot, Clarysage and other citrus oil and also with synthetics like Linalool, phenylethyl alcohol, methyl anthranillate, etc. but in all the cases are has to be taken to dose this oil, it performs well only when used in low dosage, for giving sweet pleasant affect.

It is also used in flavours particularly for pan masala, gutka and tobacco formulation. Attar is also made out of this flower which is also used in pan masala, gutka and tobacco flavour.

In aromatherapy, it is used due to its bactericidal, fungicidal, antispasmodic, anthelmintic action. It is used in fungal infection, burns, calluses and corns. Needless to say should be used in moderation. Also used as pesticide.

ECONOMICS OF CULTIVATION OF MARIGOLD

Particulars per hectare	Cost of Cultivation (Rs./ha.)
A. Cost of cultivation and Distillation	
Cost of plant material (1 kg seeds)	500
Land preparation for planting	2,000
Fertilizers	3,500
Labour cost	2,000
Irrigation	1,000
Plant protection	500
Weeding (1 time manually)	1,000
Harvesting	2,000
Transportation of herbs	1,000
Distillation charges	3,500
Miscellaneous	1,000
TOTAL	18,000
B. Gross Returns	
Oil Yield (Kg)	50 kg /ha.
Price of Oil (Per kg)	Rs. 1000/kg
Total cost of Oil (per ha.)	50,000/-
C. Net Returns	**32,000/ha./year**

The above cost parameters are based on prevailing market rate of manpower, other materials and produce (essential oil) may deviate in future.

PALMAROSA

(*Cymbopogon martini*)

Cultivation Technology of Palmarosa

Palmarosa botanically known as *Cymbopogon martini* belongs to family Poaceae, has been known for its essential oil for a long time. In India, the plant is found growing in wild state and known as Rosha grass or Rusa grass by the local people. The oil obtained from the distillation of the flowering tops is referred to as palmarosa oil in trade. There are two district varieties of palmarosa that grown side by side in wild state. The two varieties are the motia is considered superior due to its high content of geraniol (85-92%) than sofia (Geraniol Content 60-70%). The present total annual world production of palmarosa oil is estimated to be 1000 tonnes. Earlier the cultivation of palmarosa was done on a limited scale in Maharashtra, Andhra Pradesh and Tamil Nadu. In Uttar Pradesh, cultivation of palmarosa stated only in recently years. Because of rose like aroma, palmarosa oil and its major constituent, geraniol, is much valued in perfumery and cosmetic industries the world over. There has been increased demand of oil and geraniol.

Soil and climate

Palmarosa is adapted to wide range of soil climatic conditions. The plant thrives well on low fertile soils. It does not tolerate prolong aeration. Hence, soil selected for establishing palmarosa plantation should be well drained. Marginal soils, waste lands, soils affected with salinity and alkalinity problems, sandy and sandy loam soils with poor fertility status, undulated lands and soils containing stones and gravels can be utilized for economic production of palmarosa.

Land preparation

The field is prepared by one deep and two to three light ploughing. To cultivate palmarosa under irrigated conditions, beds should deceive fine leveling to avoid water stagnation. There should be provision of drainage channels to drain off excess water during rainy season. On undulated and slopy lands where plugging is not possible, plantation can be established after clearing the land of bushes and perennial weeds, and shallow tilling of soil manually. In areas with little fear of water logging during rainy season, planting is recommended on ridges (Plate 10).

Propagation and planting

It can be planted (a) by seed seedlings and (b) by splitting of clumps into slips.

(a) Planting of seedlings

It is better to select light soil near the source of irrigation for nursery. Nursery beds are well prepared and added with good amount of farmyard manure of compost. Before sowing of nursery, beds should be kept moist (at about 60% available soil moisture) for 12 to 15 days to stimulate the germination of weeds and to minimize weed competition in the nursery. It has been found that pre-sowing irrigation of nursery helps in minimizing weed competition to the extent of 70-90%. The hoeing the beds before sowing kills germinated weeds. Nursery beds are fertilized with basal dose of 25-30 kg of N, 15 kg P_2O_5 and 15 kg K_2O per hectare. Three to five kilogram seeds are mixed with 3 to 4 volumes of moist soil and are sown over a nursery area of 400-500 s cm for planting one hectare crop. After sowing, seeds are mixed in the soil by hand or by shallow tilling of the beds. Nursery beds are sprinkle irrigated for 3-4 days, thereafter beds are flood irrigated. Seedlings become ready for planting in about 35 to 40 days after sowing. Seedlings are uprooted carefully and chopped off from the top, leaving 20-25 cm from crown (ground), to avoid loss of water through transpiration. Seedlings are planted in flat beds or on ridges, maintaining 45-60 cm distance between lines and 30 cm from plant to plant. Field should be irrigated immediately after planting. In soils of low fertility and in problem soils it is desirable to maintain a closure spacing of 45x30 cm.

(ii) By splitting the clumps (Slips)

Grown up clumps is cut from 20-25 cm from ground and are dug carefully. Clumps are divided into slips containing 3-4 tillers. Slips are planted in the field, maintaining a distance recommended for planting through the seedlings. Slips give very poor establishment compared to the seedlings. Therefore, the method is useful for the multiplication of plants having high oil content and desired quality of oil for the production of seeds.

Manure & Fertilizer

Palmarosa grows luxuriantly on fertile soils; therefore, in soils of high fertility, crop does not require any fertilizer application for first two years after planting. In soils of medium and low fertility, application of 40-50 kg. Nitrogen, 30-40 kg P_2O_5 and 30-40 kg K_2O/ha is recommended before planting. Nitrogen at the rate of 20 kg/ha is top dressed 45 days after planting and 15 days after each of first and second harvests. In extremely poor soils, application of higher amount of fertilizer is needed. In red laterite soils of south India, which are low in organic matter contents, response up to 200-250 kg N/Ha has been obtained. Application of zinc has been found beneficial.

Weed control

Palmarosa is a fast growing crop and hence does not allow weeds to grow. However, it requires weed free environment for proper establishment and growth in the initial stages (up to 90 days after planting).

Irrigation

Palmarosa is cultivated both as irrigated as well as rainfed crop. Under assured water supply, the crop should be irrigated at 10-15 days interval during dry months of March-June and at 10-15 days interval during dry months of March-June and at 20-30 days interval during post rainy and winter seasons.

Diseases and Pests

Leaf blight caused by Ellisiella caudate is the most important disease. The disease appears during rainy season in almost all parts of the country. In the initial stages, grayish-brownish spots on leaf lamina appear which become enlarge in later stages of plant growth. Bernelate fungicide has been found quite effective in controlling the disease. In light soil, particularly when over aged and woody seedlings are planted, termite attack becomes very common. Therefore, in such a situation pre-plant incorporation of soil applied insecticide is recommended.

Harvesting

The crop is harvested at full bloom stage. The proper stage of harvesting comes 15-20 days after the plants start flowering. Crop planted in July becomes ready for first harvest in 3-4 months after planting in September-October. Subsequent harvests are taken in November-December, March-April and June-July. Almost the same schedule is applicable during second and subsequent year of harvests.

Distillation and Yield

Hydro-steam distillation is used for the distillation of palmarosa oil. Oil content in herb depends upon the climate harvesting season, maturity of crop and the condition of material being distilled. Plantation with very luxuriant plant growth produce herbs with significantly lower oil content, compared to plant having less growth. The average oil content in various parts of the palmarosa plant var. motia is about 0.40-0.60%. Under good management it is possible to produce 80-100 kg oil/ha in the first year, 125-150 kg in second and subsequent years of harvest.

Uses of Palmarosa Oil

The odour is sweet, terpenic, grany, rosy type. The oil is used for low cost fragrances for agarbatti, dhoopbatti, cosmetics and soaps. The Palmarosa oil is very good for making floral, sweet and other fancy fragrances.

Also find use in flavours for tobacco, pan masala, sweet supari and khaini.

The oil is used for treatment of face, hands, feet and lips, moisturizing skin, good for acne and wrinkles. It also stimulates cellular regeneration, regulates sebum and effective in neurologic and rheumatic pains.

The geraniol obtained out of Palmarosa oil is of the finest quality and highly priced in fragrance & Flavour industry and is used for making good quality rose and other floral fragrances. This geraniol ex Palmarosa is used in flavours as well particularly rose, vetiver etc.

ECONOMICS OF CULTIVATION OF PALMAROSA

Particulars per hectare	Cost of Cultivation 1st year (Rs.)	Cost of Cultivation 2nd year (Rs.)
A. Cost of cultivation and Distillation		
Cost of plant material **(4kg seeds)**	4,000	----
Land preparation for planting of seeds	2,000	----
Fertilizers	2,000	2,000
Labour cost	2,000	2,000
Irrigation	2,000	5,000
Plant protection	1,000	1,000
Weeding (2 times manually)	2,000	2,000
Harvesting	1,000	2,000
Transportation of herbs	1,000	2,000
Distillation charges	2,500	3,500
Miscellaneous	500	500
TOTAL	20,000	20,000
B. Gross Returns		
Oil Yield (Kg)	80 kg/ha.	100 kg/ha.
Oil Price (Per kg)	Rs. 700/kg	700/kg
Cost of oil produced (per ha.)	56,000/-	70,000/-
C. Net Returns	36,000/ha./ year	50,000/ha. /year

The above cost parameters are based on prevailing market rate of manpower, other materials and produce (Essential oils) may deviate in future.

PATCHOULI

(*Pogostemon patchouli*)

Cultivation Technology of Patchouli

The Patchouli is an important aromatic plant cultivated for its essential oil which is obtained by steam distillation of matured dried leaves. The botanical name of patchouli is *Pogostemon patchouli* belongs to family labiatae. The oil is known to blends well with other essential oils, sandalwood, geranium and lavender. Patchouli oil is considered to be the cheapest fixative among the natural source, and enhances the life of any perfume to which it is part. Presently, there is no synthetic substitute of patchouli oil so it enhances the value and unique position of patchouli oil in perfumery (fragrances and flavours) and cosmetic industry (Plate 11).

The old literatures reveals, the word "Patchouli" was derived form Sanskrit world "Patchouli". In ancient times, Indian fabrics and shawls, prepared for export to European countries, were permitted with strong odour which was considered characteristics and proof of oriental origin. This was only in about 1844 when the first shipment of dried patchouli leaves arrived in London and the long sought secret of mysterious scent could be revealed. Thereafter, French manufacturers learned and started using patchouli leaves to their costly shawls in close intimation to the East Indian fabrics.

The plant of Patchouli was first described in 1845 by Pelletier Sautelet, and named *Pogostemon patchouli.* For many years, the distillation of patchouli oil was confined to British-Malaya, mainly on the Iceland of Penang and Singapore. In the beginning of 20th century, the production of patchouli oil took a new turn when plantings were developed in the province of Sumatra, Indonesia. Initially, the cultivated leaf material was exported to European countries. This practice continued until before the Second World War, when patchouli industry in Sumatra developed rapidly and Indonesia soon stated to produce more than 85% of total world production. Now-a-days Indonesia also meets more than 80% of total world requirement of patchouli oil. The remaining 20% is produced by China and other countries with very limited production in Malaysia, Brazil and India. The total world production of patchouli oil is estimated to be more than 800 tonnes annually, of which about 650-700 tonnes is produced in Indonesia. About 80-100 tonnes oil worth Rs. 2.2-2.8 crores is imported annually in the country. Therefore, there is need to popularize the cultivation in different parts of the country to make India self-sufficient in patchouli oil production in the open areas of coastal belt and orchards in Northern Indian plains.

Soils and Climate

The plants of patchouli can be cultivated in a wide range of soil and climatic conditions. However, the plant flourishes best in hot and humid climate with evenly distributed rainfall. The plant can be cultivated on undulated land and terraces where there are remote chances of water logging. The successful cultivation also possible in plains at low altitude provided proper drainage of the field is ensured by planting on ridges and laying out proper water drainage channels. For best growth plant requires moderate temperature (25-30^{0}C) and high humidity ranges 60-70%. Partial shady conditions are considered for good growth. Hence, the spaces between the mango trees in the orchards of different kinds can be considered suitable for its cultivation. Although, patchouli does not removes much of the nutrients from the soils, availability of reasonably good amount of nutrient in the soils considered essential for optimum productivity. The above fact holds true considering very luxuriant growth and high productivity obtained from the plantations established on virgin forest lands rich in organic matter. Patchouli plantation requires much attention in subtropical climates, particularly during hot and dry summer in the months of May and June.

Land preparation

Before plantation, land is cleared off of bushes and perennial weeds through deep ploughing and digging the spots to uproot the deep rooted bushes. Individual beds are finely leveled to avoid water stagnation and to ensure proper drainage. In areas with little fear of water logging, high ridges are made to save plants especially in the month of July-August.

Propagation

It is vegetatively propagated especially by stem cuttings from older plants. For best results apical cuttings 10-15 cm long with 3-4 leaves are taken from well developed branches. These cutting are either used for raising nursery or can be planted directly in the final field. For raising nursery, cuttings are planted in bags or nursery beds containing sufficient sand and leaf-mould mixture, etc. Nursery is raised in the shady place with provision of good drainage. Before planting the cuttings in nursery or in seed beds, these are treated with root promoting hormones for quick establishment and fast growth. The cuttings are stripped off of one or two fully developed lower leaf before planting. Nursery beds are kept moist for 8-10 days, till the rooting starts, through sprinkler irrigation 2-3 times daily. Best times for raising nursery is when the temperature is mild (20-25^{0}C) and humidity is around (70-80%). About 4-5 weeks time, nursery plant becomes ready for planting in field. Mist chambers are best suited for propagating patchouli.

Planting

In subtropical plains, planting should be preferred on ridges and furrows and for direct planting, it is advisable to start the planting operation when the weather is cloudy and likely to be rained. The cuttings or nursery raised plants are planted on the spacing of 70-85 cm apart at 65-70 cm plant to plant distance. Before preparation of ridges, an application of 20 kg. Furadan or 150 kg. neem cake is advisable. If the planting is done in open field desired shade can be creating, otherwise, in open field desired shade can be created by planting trees like, popular, mango, teak, etc. few months before planting of patchouli. In any case the newly planted cuttings or nursery plants should be protected against hot sunlight. Irrigation of field is immediately required just after planting of patchouli.

Manures and Fertilizers

In forest land or in orchards which are rich in organic matter, and patchouli plantation is required to be established for the first time, only a basal application of 30 kg nitrogen, 60 kg P_2O_5 and 35 kg K_2O per is considered sufficient for one ha. of land. However, when the plantation is done on a medium fertile soil, application of 40 kg each of N, P_2O_5 and K_2O per ha is applied as a basal doses and 100-125 kg N is applied in 3 equal splits doses after each cutting. Use of phosphorus has been found to be quite responsive to patchouli, hence, soils different in phosphorous may need higher application of phosphate fertilizers.

Irrigation

The availability of sufficient soil moisture throughout the growing period is the key factor for successful production of patchouli plants. Therefore, areas enjoying well distributed rainfall and long rainy season are well suited for patchouli cultivation. During dry months frequent irrigations are required to keep the soil moist. In sub-tropical climates during dry summer months, March-June, irrigations at 8-15 days intervals are required.

Diseases, Pests and their control

The yellowing, leaf blight disease and root knot nematode are the important diseases and insect pests of patchouli. Root knot caused by Nematodes, is most serious and destructive pest of patchouli. Infected plant roots develop knots varying from 0.4 - 4.5 cm in diameter. In the initial stages of infection, it is difficult to differentiate the infected plants with normal plant. However, in advanced stages, plant shows stunting of top growth and starts wilting. In absence of any control measure, the nematode spreads at an alarming rate and plant starts drying. Within 1.5-2.5 months the entire crop stand died and perished. Application of suitable nematicide should be incorporated in the soil and follow a light irrigation. Precautionary measure should include regular application of Neem cake, crop rotation with grass species and intercropping of nematode resistant crops is preferred. Leaf blight disease is caused by fungus, leaves bear brown spots near the margins which enlarge into patches and causes death of leaves. The disease is controlled by spraying of suitable fungicide like Dithane M-45 or any copper fungicide. Another disease is caused by virus, molting of leaves rolling and crinkling are the important symptoms. As soon as the symptoms appear, infected plants should be uprooted and burnt separately.

Harvesting

After first cutting, crop is harvested at about 3-4 months interval upto 2-3 years. At the proper stage of harvest the plant growth become slow, the foliage starts turning pale green and leaves emit typical patchouli odour. Harvesting is done with sharp sickle/knife in the early morning or in the evening and not during hot noon hours. All the young growing leaves and tender branches 35-45 cm above ground are cut. The cutting close to the ground may lead to considerable delay in re-generation and a long gap between harvest, thus, affecting the total oil production. Patchouli plantation remains productive for 3-3.5 years, after which the plantation is uprooted and rotated with nematode resistant crops for few years.

Drying and Curing of leaves

The harvested/plucked leaves and tender twigs is spread out to drying in thin layers on concrete floor. Drying can also be done on racks made of bamboos. In any case it should ensure that complete drying is done in 4-5 days time. Proper drying is of great importance for obtaining good quality oil. During whole drying period herbs should be frequently turn over by hand or any other tools to ensure even and through drying and to prevent fermentation. In absence of drying shed, having free air circulations, sun drying is practiced which caused some loses of oil and its odour value. Properly dried leaves develop a strong and characteristic patchouli note which is less noticeable in fresh leaves. To avoid the chances of fermentation of leaves, it is advisable to harvest the crop when the sky is clear and the atmospheric humidity is low. The highest oil yield was found

(5.0-6.0%) in 140-150 days dried leaves. In case of delay in distillation, the dried leaves should be pressed and stored in dry place. Otherwise, the leaves stored loosely, especially during rainy season, may develop mouldy odour due to high atmospheric humidity.

Oil content and yield

The dried leaves contain about 4.0-5.0% whereas the tender stem has only 0.5- 0.8% essential oil. Depending upon the quality of herb and the method of drying the oil recovery on commercial scale distillation varies from 2.5-4.0%. The average yield of fresh herb is about 125-150 q/ ha. is harvested form 3-4 harvests in a year, which on drying remains to 30-32 quintals and yields about 60-70 kg. essential oil. The herb and oil yields are greatly affected by the age of trees in the orchard area occupied by trees and the amount of sun light available to the patchouli plants. In highly shady conditions, the herb yield and oil content is considerably reduced as the branching, leaf development and oil biosynthesis in the patchouli leaves are adversely affected. The 60-65% shade in orchards is suitable for patchouli plantation.

Uses of Patchouli Oil

The odor is sweet herbaceous wood balsamic with rooty undertone. An ethereal wine like sweet floral note makes it quite useful in fine fragrances. This ethereal top note is little rare. Regarding flavour it is spicy woody and bitter.

It is used a lot in fragrances to its tenacious nature used in fragrances for all types though little in floral. It is stable if used in fragrances for soap. It blends well with sandalwood, vetiver, labdanum, granium, lavender, clarysage, clove, bergamot and neroli. It also blends well with synthetic loke ionones, coumarin, nito-musk, methyl salicylate and cinnamates. It is used extensively in making oriental, fougere, chypre and woody fragrances.

It is used for making the odor of depilatories particularly when used with citrus oils. Also used for making agent for alcoholic breath, onion/garlic odor, in after dinner candy. Such flavour are made when used with geranium, ionone, orris, anis, clove. It is used in flavour for mouth freshners, tobacco, very extensively in pan masala and gutka flavour. Also used in baked goods, candy and in chewing gum.

It is used in aromatherapy a lot owing to its action as anti-depressant, anti-inflammatory, anti-microbial, antiseptic, antiviral, astringent, stimulant, bactericidal, etc. used in skin care for acne and chapped skin, oily skin. For hair care it is used for dandruff, oily hair. Also good for fungal infection, dermatitis, wrinkles. Good in case of repelling moths. Effective in nervous exhaustion, stress etc.

ECONOMICS OF CULTIVATION OF PATCHOULI

Particulars per hectare	Cost of Cultivation 1st year (Rs.)	Cost of Cultivation 2nd year (Rs.)
A. Cost of Cultivation and Distillation		
Cost of plant material (17,000 cutting @ Rs.	17,000	----
1/cutting)	2,500	----
Expenditure on nursery preparation	2,000	----
Land preparation for planting	2,500	2,500
Fertilizers	2,000	2,000
Labour cost	2,000	3,000
Irrigation	1,500	1,500
Plant protection	2,000	3,000
Weeding (2 times manually)	4,000	5,000
Harvesting/Picking of leaves/Drying	1,000	2,000
Transportation of leaves/herbs	4,500	5,500
Distillation charges	500	500
Miscellaneous		
TOTAL	41,500	25,000
B. Gross Returns		
Oil Yield (Kg)	50 kg/ha.	60 kg/ha.
Oil Price (Per kg)	Rs. 1500/kg	1500/kg
Cost of oil produced (per ha.)	75,000/-	90,000/-
C. Net Returns	33,500/ha./year	65,000/ha./year

The above cost parameters are based on prevailing market rate of manpower, other materials and produce (Essential oils) may deviate in future. The returns will be same from 2nd & 3rd year.

ROSE

(*Rosa damascena*)

Cultivation Technology of Rose

Rose is an important aromatic crops cultivated for its essential oil obtained from hydro and steam distillation of flowers of *Rosa damascena* belongs to family Rosaceae. This is one of the costliest and sweetest fragrant materials known to the world from the ancient times. It is a perennial plant with an average life span of 20-25 year. Certain species can survive up to 50 years. Rose oil has very wide application for making costly fragrances. Perfumery, cosmetics and tobacco industries are the major consumers of rose oil, on limited scale it is also used for flavouring purposes (Plate 12).

In India rose cultivation is done onabout 10,000-12,000 hectares of land. The important center of rose cultivation are Hasayan (Hathras), Kannauj and Etah in Uttar Pradesh and Pushkar (Ajmer) and Haldighati in Rajasthan. The Damask rose, (*R. damascena*) cultivation is confined to Hasayan, where it is grown on approximately 5000 hectares of land. Here, the major flowers produced are used for production of rose oil, rose water and attar.

A species of rose i.e. *R chinensis* (Adbourd) which is also called cheenia gulab locally, as it continues to flower throughout the year. Flowers at these places are mainly processed for the production of attars, rose water, gulkand and dry petals. Damask rose is called "Chatti" or "Phasali Gulab" as the main flowering season of *R. damascena* is March-April, however it flowers during October-November also.

Soil and Climate

Although, rose can be cultivated in a variety of soils, medium loam well drained soils are considered most suitable. Acidic soils are avoided for its cultivation but slightly alkaline soil (pH 8.0-8.5) may be used for establishing rose plantation. Land infested with perennial weeds may be avoided. Rose plantation can also be established on sloppy land and terraces in hilly areas, provided irrigation system other than flood irrigation is available for watering the plantation.

Rose is basically a temperate plant and requires mild temperature for optimum growth. Considering the above facts, hilly areas and mild hill conditions are most suitable for cultivation. Rose can also be cultivated in sub-tropical climate, however, the yield in sub-tropical climate is lower than temperate and sub-temperate climate. In India, most of rose growing areas falls in sub-tropical climate, however, the yield in sub-tropical climate is lower than temperate and sub temperate climate. In India, most of rose growing areas falls in sub-tropical climate. Recently rose cultivation has been introduced in temperate climate of J & K and Himachal Pradesh from where the level of yield reported is much higher than sub-tropical. Bright sunny days, high humidity (60%) with mild temperature (20-25° C) is most conducive for enhancing the flower production, as these conditions prolong the duration of flowering and give synchronized flowering.

Land Preparation

Field for rose plantation should be deep ploughed several times to ensure the removal of deep rooted perennial weeds, deep summer ploughing can be helpful in eradicating perennial weeds. With subsequent 3-4 light ploughing the soil is brought to a fine tilth. The field is divided into beds of convenient size Individual beds are leveled with great care to help uniform distribution of irrigation water in the beds.

Propagation

Rose is propagated vegetative be means of stem cuttings. Cuttings can be planted directly in the field or raised in nursery. Further, for direct planting there are two methods, commonly prevalent among growers. In first method 25-30 cm. wide cuttings are made 2-2.5 m apart, end-to-end. Cuttings are covered with mixture of soil and well rotten farm yard manure. In second method, cuttings are planted in pits of 30x30x30 cm. Four cuttings of 20-25 cm in length are planted in the centre of each pit vertically with upright position. At least 1/3 portion of the cutting is dipped into the soil. The soil around cuttings is pressed properly. The first method is more common among the Bulgarian farmers while the other with rose growers in India.

For raising nursery, bundles of 50-100 cuttings are made and these bundles are placed in pits in upright position and covered with soil. The soil around the cuttings is kept moist until they are taken out for planting in nursery. After 20-25 days the cuttings become ready for planting in the nursery. During this period the cuttings are full of callus formation and buds are sprouted. Cuttings in the nursery beds are planted in rows 20-30 cm apart at 15 cm distance between the cuttings. The cutting remains in nursery beds until July-August when these are shifted in the main field. Irrespective of methods of planting, it is advisable to treat the cuttings with any root growth promoting hormones for early rooting. The optimum time for taking cuttings is December when roses are pruned.

Planting

Two methods of direct planting have already been described above. For planting nursery-raised cuttings pit of 30x30x30 cm or 45x45x45 cm are made at 1x1 m spacing. In temperate regions the row to row and plant to plant distance may be kept at 2x2 m or even 2.5x2.5 m pits are filled with soil mixed with well rotten farmyard manure (FYM) at the rate of 15 kg. per pit. Aldicarb 20% at the rate of 2-3 g per pit is added to the soil and F.Y.M mixture to protect the plant from termite attack. The plants from nursery are dug out carefully and planted in the pits. The optimum time for planting in subtropical regions is July-August whereas for temperate regions October- November or February-March.

Manures and Fertilizers

Rose grows well in soil rich in organic matter and has high nutrient status. Apart from farmyard manure (10-15 tones per ha) that is added in the trenches or in pits, fertilizer application is done to supplement the nutrient requirement. A dose of 100 kg nitrogen, 50 kg. phosphorous and 50 kg potassium is applied at the time of planting. Nitrogen and the potassium is applied at the time of planting. Nitrogen at the rate of 50 kg. per ha is top dressed 30-45 days after pruning. In subsequent years the above manure and fertilizer application schedule is followed. In second year onward farmyard manure and basal dose of fertilizer in application schedule is followed. In second year onward farmyard manure and basal dose of fertilizer is applied after pruning. The schedule of nitrogen top dressing remains the same for planting as well for subsequent years of plantation.

Weed Control

Rose is susceptible to weed interference mainly in the planting year. Therefore, special care is taken to suppress the weed growth for about 6-7 months after planning. To manage weeds during the period 2-3 manual weeding are required. In second and subsequent years the plants develop sufficient canopy and take care of weeds. In established plantation, one hoeing, after pruning in December, and a weeding in January-February may be sufficient to keep the plantation free of weeds.

Irrigation

For better plant growth and flower production rose requires frequent irrigations. During bud formation and flowering and also in dry months plantation should be irrigated regularly.

Pruning

Pruning of rose plants is an important operation. The pruning operation is carried out in the first fortnight of December in Plains (sub-tropical) and October-November in hilly areas (temperate). Both, early and delayed pruning have very adverse effect on flower production. In sub-tropical climate heavy pruning is practiced where all the branches 20-25 cm above the ground are removed, using a sharp tool. In temperate region light pruning is done and branches 75-100 cm above the ground are removed. After 10-12 years of planting the flower yield starts to decline. Therefore, it becomes desirable to rejuvenate the plantation. In this process bushes are cut from 8-10 cm above ground and the soil is dug to a depth of 20-25 cm Farmyard manure 15-20 tonnes per ha and mixture of N.P.K. fertilizers is applied. After irrigation the plants start producing new shoots.

Diseases, Pests and their control

The most important disease of *R damascena* is black spot caused by the fungus (*Diplocarpon rosae*). Dark brown or black spots start appearing on infected plants. In advance stage plant growth is checked and defoliation takes place. Dieback is another disease of rose caused by fungus, *Diplodia rosarum*. A symptom of disease starts appearing after the pruning operation when cut ends start turning black. As a preventive measure the plantation is sprayed with copper fungicide immediately after pruning. Another spraying with fungicide Bavistin should be done a month after the pruning when the foliage development has taken place.

Picking of Flowers

Flowering in rose plantation commences two years after planting. However, the full flower production can be expected in third year onward. The flowers are picked daily early in morning so as to complete the operation before the sunrise. Depending upon the soil and management practices flower yield in sub-tropical climate from third year onward ranges from 2000-3000 kg/ha/year. In temperate climate flower yield ranges form 4000-5000 kg/ha/year. Flowering in sub-tropical climate starts from middle of March and ends up by 3rd week of April. In temperate climates, blossom starts in May and continues for about 50 days.

Distillation

As soon as the harvesting is over flowers are taken to distillation plant for processing. In case, there is delay in processing, flowers should be spread in shade at low temperature and water should be sprinkled to avoid the loss of oil through evaporation. In India, three different methods are in practiced for the processing of rose oil.

(i) Deg and Bhapka Method

Majority of rose distillers in India use traditional distillation equipment which consists of round kettle (*deg*) and a receiver (*bhapka*), which also serve the purpose of condenser. These are made of copper. The kettle and the receiver are connected with bamboo (*chonga*) through which steam from kettle reaches to the receiver. The receiver is placed in a small water tank which helps in the condensation of vapours containing rose oil. The kettle has a capacity of 50-100 kg flowers per batch and takes about 6-8 hours to complete the distillation. After distillation is over the receiver is removed and oil is separated. This method gives a recovery of 0.01-0.015% which is significantly lower than the other methods described below.

(ii) Direct Fired Distillation Unit

The unit consists of a kettle (still), a column, condenser and receiver. The whole unit can be of copper or stainless steel. The capacity of still may range from 100-250 kg. flowers per batch. To increase the fuel efficiency and to generate sufficient steam, fume tubes are fitted in the bottom of the still. The outlet of the receiver is connected with the column to recycle the condensate into the still, after the separation of oil takes place in the receiver. This process in known as "Cohobation" which has been found to improve the recovery of oil. The plant is simple in operation and can be operated by a semi skilled worker. Direct fired plant may cost Rs. 1.25-1.5 lack, if made out of stainless steel. The plant takes about 4-6 hrs per batch for completion of distillation.

(iii) Boiler Operated Unit

This is most modern method of distillation and suitable for large-scale production of oil. The whole plant consists of three units, a boiler, a distillation unit and a "Cohobation" unit. Flowers are charged in the distillation unit and distillate after passing through a receiver fed to another still through cohobation column to ensure complete recovery of oil. Part of the oil is also collected from the first receiver and the remaining from the second. It is advantageous to collect the distillate from the second separator and feed back it to the distillation tank in pace of fresh water. Complete distillation in boiler operated distillation unit takes 4-5 hrs. The oil recovery varies from 0.025% -0.03%. Depending upon the capacity; of the boiler, the unit may cost Rs 6-8 lacs. The other advantage with the system is that several distillation units can be connected with the boiler.

Oil content and Yield

Although scattered flowering may be noticed in the planting year itself, optimum flower yield is obtained in third and subsequent years of plantation. This yield level is maintained until ten years of plantation, thereafter tends to decline. In subtropical climate an average flower yield of 200-300 kg in the first year, 500-800 kg in the second year and 200-300 kg. /ha. In third and subsequent years of planting is obtained. Thus, with an average 0.020% recovery, oil yield during third and subsequent years may vary from 0.4 to 0.60 kg/ha/year. Considering an average recovery of 0.03% oil in temperate climate, the oil yield in third and subsequent year may range from 1.20 kg to 1.50 kg/ha/year from 4000 to 5000 kg flowers/ha harvested each year.

Uses of Rose Oil

The odour is floral rosy, deep rich as if rose flower are present in bulk. This oil finds use in high quality fine fragrances. It is also used in pan masala and tobacco blends.

It gives a soft natural character in food flavours as well. These materials also find use in flavours in trace amount in apricot, strawberry, raspberry, bitter almond, apple, etc. Flavour is sharp at high conc, but turns to be soft, warm, sweet, balsamic upon extreme dilution.

The oil is also useful for dry, natural and sensitive skin. It is also effective for wrinkles, eczema and as a cardiac tonic.

Rose flower can be used for making gulkand which is an excellent laxative. The flower also finds its use for decorative, ornamental purposes and in gift as a bouquet.

Rose water is used in making softdrink, flavouring sweets. It acts as a skin toner, hence, used in face packs and other herbal cosmetic products.

Rose absolute which is produced out of rose flower is produced extensively in high and medium price perfumes particularly in floral bases, chypres, oriental bases, etc. They are also used to round off the sharp notes in synthetic composition. It blends well with Jasmine, Cassie, Mimosa, Orange flower and other floral as well as with most of the synthetic perfume materials.

The rose oil and absolute both are highly diffusive in nature and because of this character they form a good place in high to medium price perfumes.

The waxes produced during the rose concrete also find use in soaps. The rose concretes are also used for perfuming the solid perfume waxes.

ECONOMICS OF CULTIVATION OF ROSE

Particulars per hectare	Cost of Cultivation 1st year (Rs.)	Cost of Cultivation 2nd year (Rs.)
A. Cost of Cultivation and Distillation		
Cost of plant material (10,000 plants)	50,000	----
Land preparation for planting of seeds	5,000	----
Fertilizers	4,000	4,000
Labour cost	10,000	10,000
Irrigation	5,000	6,000
Plant protection	1,000	1,000
Weeding (2 times manually)	2,000	2,000
Harvesting	2,000	2,000
Transportation of herbs	2,000	2,000
Distillation charges	8,000	16,000
Miscellaneous	1,000	1,000
TOTAL	90,000	44,000
B. Gross Returns		
Oil Yield (Kg)	0.50 kg/ha.	1.0 kg/ha.
Oil Price (Per kg)	Rs. 2,50,000/kg	2,50,000/kg
Cost of oil produced (per ha.)	1,25,000/-	2,50,000/-
C. Net Returns	35,000/ha./ year	2,06,000/ha./ year

The above cost parameters are based on prevailing market rate of manpower, other materials and produce (Essential oils) may deviate in future.

ROSEMARY

(*Rosmarinus officinalis*)

Cultivation Technology of Rosemary

Rosemary is botanically known as *Rosmarinus officinalis* is extensively used in food, flavour, and fragrance industries. The essential oil of rosemary is mainly used in the production of soaps, detergents, freshener sprays, as an antioxidant to improve stability of foods and as a fixative and blends well with several other essential oils. The oil has anti-malarial and insecticidal properties. It can be effectively used as mosquito repellent (killer as it can kill about 90 to 100 percent of adult mosquitoes and larvae. Rosemary oil is also used as vascular smooth muscle relaxant, flavouring agent in edibles and in distilled water for eyewash. The dried leaves of rosemary find use in culinary items. (Plate 13)

Rosemary is cultivated in large scale in Spain, Italy, France, Algeria and Portugal. Morocco is leading producer of Rosemary oil in the world and world production is about 75 -100 tonnes per year. The plant is cultivated in cooler climate of Nilgiris in south India and planes of Karnataka. It thrives well at elevation of 1000-1500 meter above sea level. The crop can also be cultivated in lower reaches of Himalayan States of Jammu and Kashmir and hills of Uttarakhand, Himachal Pradesh, northern hills of West Bengal and Arunachal Pradesh.

The plant of *Rosmarinus officinalis* belonging to Lamiaceae family, is a dense evergreen shrub of 60 to 75 cm height with woody branches bearing cylindrical inwardly folded 2-4 long thick, sessile leathery leaves having characteristic, aroma and green and rough above and white hairy below. The flowers are white with bluish tinge arranged at the end of the branches in little clusters.

Soil and Climate

It grows well on drained red clay and sandy loam soils in cooler places where temperature does not exceed 30°C. In hilly areas the planting is done in December/January and in planes the planting period is March and April.

Propagation

The plant is propagated vegetatively by terminal stem cuttings of about one year old plant. Apical cuttings of 8-10 cm length are suitable for multiplication. Upper leaves are kept intact and lower leaves are removed from the cuttings before planting. The cuttings are planted in a mixture of soil, sand and leaf mould in polythene beds, which are kept in nursery in the shade and regularly moistened with water twice a day. Cuttings sprout after about 40 days and ready for planting. The treatment of cutting with 5000 ppm. naphthalene acetic acid (NAA) increases the rooting percentage. The best period for planting is February /March in north India and October-November in south India.

Transplanting

First of all the land is ploughed to a fine tilt and laid into beds of 3 x 3 m and 10 tonnes of FYM is applied. Subsequently,40 kg of superphosphate and 66 kg of muriate of potash are applied to the soil. Rooted cuttings are planted in rows at spacing of 60 x 45 cm. The beds are irrigated lightly after two weeks of transplantation.

Irrigation

Initially, 1-2 irrigations per week are required to establish the crop and subsequently weekly irrigations are required. It is totally depend on type of soil and climate.

Disease, Pest and their control

Rosemary is adversely affected by *Rhizoctonia solani,* a soil borne fungus causing severe leaf blight, which is controlled by fumigate spray of Benomyl (0.1%) or Mancozeb (0.3%).

Harvesting & Yield

Rosemary is harvested with the on set of flowering tops measuring 30-50 cm along with leaves and cut with sickle. The woody shoots should not be used for distillation of oil as they give unpleasant odour of turpentine. In the first year the crop is ready after 4 months of planting and in subsequent three years, 3 harvests are obtained at intervals of 4 months/year. About 12 to 15 tonnes of green herbage is obtained in a year from one hectare of land, which gives 84 to 100 kg of oil/ha/year. As such yield of oil is 0.8-09%%. The essential oil of rosemary is colourless to light greenish yellow with camphoraceous odour imparting penetrating and sharp top note. The major constituents of rosemary oil are L-pinene, 1,8 cineole camphor and linalool.

Distillation

Essential oil of Rosemary is obtained by steam distillation of leaves and twigs, which are shade dried before distillation of essential oil. Distillation takes 3-4 hours for full recovery of oil.

Uses of Rosemary Oil

The odour is strong fresh, herbaceous, balsmic in nature. The oil is used for fine fragrances and also for fragrances for soap & cosmetic products and room freshener. It is used in making colognes, lavender water, fougeres, pine needle fragrances, oriental perfumes, in room deodorants, house hold sprays as insecticide, disinfectants, etc. This is also used in fragrances for agarbatti and dhoopbatti and also for low cost detergents. The oil is very good when a herbal note is desired in a fragrance.

Due to its low boiling and fresh smelling odor it has got a good effect as a masking agent particularly for phenolic and tar like odors. The terpenless rosemary oil is used in alcoholic perfumery. The rosemary oils blends well with lavender, citronella, labdanum olibanum, cedarwood oils and finds its application in making lotions as well.

The oil is an excellent antioxidant, anti dandruff, anti bacterial and anti fungal as well. Hence, it is used in a number of hair care and skin care products.

ECONOMICS OF CULTIVATION OF ROSEMARY

Particulars per hectare	Cost of Cultivation 1st year (Rs.)	Cost of Cultivation 2nd year (Rs.)
A. Cost of Cultivation and Distillation		
Cost of seeds/plant material	8,500	----
Land preparation for planting of seeds	2,000	----
Fertilizers	2,000	1,000
Labour cost	4,000	4,000
Irrigation (6)	4,000	5,000
Plant protection	500	500
Weeding (2 times manually)	2,000	2,000
Harvesting	2,000	4,000
Transportation of herbs	2,000	2,000
Distillation charges (5 batches) @ Rs.	7,500	9,500
Miscellaneous	500	500
TOTAL	35,000	28,500
B. Gross Returns		
Oil Yield (Kg)	80 kg/ha.	90 kg/ha.
Oil Price (Per kg)*	Rs.1,500/kg	1,500/kg
Cost of oil produced (per ha.)	1,20,000/-	1,35,000/-
C. Net Returns	85,000/ha./year	1,06,000/ha./year

The above cost parameters are based on prevailing market rate of manpower, other materials and produce (Essential oils) may deviate in future.

TUBEROSE

(*Polianthes tuberosa*)

Cultivation Technology of Tuberose

Tuberose is commonly called as *Rajnigandha,* cultivated for its white tube like fragrant flower, concrete and absolute. The tuberose belongs to the genus *Polianthes* and species *tuberose,* the name 'tuberose' is derived from *tuberosa,* this plant being the tuberous hyacinth. Tuberose was cherished and cultivated in Mexico before the Spain, it is one of the many flowers which have come to us from the ancient culture. The two species i.e. *Hyacinthus Indicus tuberosus flore,* called 'tuberose' and *Hyacinthus Indicus tuberosus flore pleno* commonly called 'the double tuberose' found in the world. This plant was taken to Europe towards the end of 16th century. How and when the tuberose found its way to India and later to Sri Lanka and elsewhere in the orient is probably an un-documented history. (Plate 14)

Soil and Climate

Tuberose can be cultivated in various types of soils preferably pH range of 6.5 to 7.5, light sandy loam to a clay loam. At least 50 cm deep, well drained, friable soils rich in organic matter and nutrients with plenty of moisture are preferable. It can be grown successfully as a commercial crop under high saline-alkaline conditions with better agronomical practices, some scientists observed that increasing the levels of NaCI and very low concentration of $CaCl_2$ affects vegetative growth and flowering of tuberose. Tuberose can be grown with success under wide environmental conditions ranging from tropical to sub-tropical and temperature climate. If the climate is mild and free from extremes of high and low temperature, profuse flowering can be obtained almost throughout the whole year. Tuberose should be planted when the temperature is between 24° to 30°C; for plant growth 13 to 18°C is good, and the blooming is better from 10 to 13°C. Commercial cultivation of tuberose in India is mainly confined to warm humid areas with average temperature range from 20 to 35°C. High atmospheric humidity and temperature around 30°C is optimum. The Spike length and quality of flower is affected if the temperature is above 40°C; very low temperature and frost also damage the plants and flowers.

Propagation

Tuberose is propagated by both methods from seeds, bulbs and bulblets. The seeds are difficult to germinate and hence, these are to be sown evenly at a depth of 1.0 to 1.75 cm in well prepared mixture containing leaf mould, garden soil and sand in equal proportions in nursery bed or seed pan. Under favourable soil temperature (26.6°C) and moisture, the seeds germinate within two to three weeks after sowing. When the seedlings are sufficiently grown, they are transferred to the final field. Multiplication of tuberose is done commercially by bulbs. Tuberose bulbs planted once can be kept undisturbed for 2 to 3 years to come. If large sized bulbs are planted and proper care is taken the flower yield will be maximum in first two years, and thereafter the yield will reduce successively in the following years. Number of spike per plant varies from 1 to 3, which attain a length of 90 to 120 cm with 20 to 35 flowers. Tuberose bulbs are easily winter killed, so in temperate climate, they are usually dug and stored during the winter months. The small new bulbs require three years of cultivation before they are capable of producing bloom.

Preparation of Land

Before planting of tuberose, first ploughing is done in the month of January so that, in the dry season with good sunshine, the uprooted ploughed weeds are completely destroyed. Second ploughing is done about a month before planting, and well rotted farm yard manure or cow dung be incorporated during ploughing operation at the rate of 40-45 tonnes per hectare. Then the clods are to be loosened, soil is brought in good tilth and the weeds are thoroughly removed. Tuberose is a sun-loving plant and bloom luxuriantly on well drained soils possessing sufficient moisture holding capacity. They are very susceptible to poor drainage conditions.

Planting

Planting of fresh bulbs as such leads to profuse vegetative growth and less flowering. Before planting, it is better to store the bulbs for nearly two months in a well ventilated warm (minimum temperature 12-14°C) dry place. If kept moist and cool, the bulbs are likely to rot at the centre. Healthy tubers will always be green at top or show some sign of life at the growing point. Before planting, bulbs should be treated with copper fungicide. In India, tuberose is generally planted in February-March in plains and in April-May in hills however, in south India, planting of bulbs during July-August is preferred. To obtain flower almost throughout the year, sequential planting can be practiced. Planting of large size bulb (3.0 to 3.5 cm in diameter) significantly improves the yield and quality of flower spike. It is planted at 3-10 cm depth, depending upon size of bulb and texture of soil viz., deeper in light soils. Spacing varies with the soil and climatic conditions, 1,00,000 to 2,50,000 bulbs are required for planting in one hectare of land with the distance of 20 cm x 30 cm and 30x 30 cm spacing is recommended according to soil condition.

Manures and Fertilizers

The nutritional requirement of tuberose varies with the soil and climate, in normal soil, FYM at the rate of 25 to 45 tonnes per hectare help proper establishment of bulbs. Apart from the organic manure, Nitrogen, phosphorus and potassium markedly influence the growth and development of plants. The deficiency of nitrogen and phosphorus completely suppressed flowering in tuberose. The deficiency of calcium causes cracking of the flower spike and its acute deficiency results in bud rot. Magnesium deficiency causes leaf bronzing, promoting yellowing on both sides of the midrib and poor growth. Sulphur is required to increase oil content in flowers, and its deficiency causes chlorosis of leaves. Iron, manganese, zinc, aluminium and copper are available at low pH ranges, whereas the availability of boron decreases below 4 and above 7.5 pH ranges. In alkaline-saline soils, application of 35 kg nitrogen and 55 kg phosphorus per hectare has been found very beneficial and further analysis revealed that very high quantities of salts are removed by tuberose. However, innormal soil 100 kg Nitrogen, 50 kg phosphorous and 35 kg potash is required for good growth and yield of flowers.

Inter-culture and Aftercare:

Proper weeding, hoeing and irrigation of crop resulted for good growth and yield. The earthling up enables the spikes to grow erect and hoeing in the plantation is routinely done to soil for attaining good growth. About 3-4 manual weeding is recommended to avoid growth of unwanted weeds. Regular irrigation is required for good growth and yield of the crop. First irrigation is applied just after planting of bulbs and rest is as and when required.

Pests and Diseases

To avoid pest and disease free field regular watch of the crop is required. Tuberose is a hardy crop and are less affected by the attack of insects, pests and diseases. However, aphids and grasshopper feed on the flower buds and young leaves which can be controlled by 0.1 per cent spray of malathion or rogor at fortnightly interval. Red spider mites suck plant sap, and the leaves become yellow or bronze in colour. Few other common pest are attached on the plant, Red spider mites suck plant sap, and it is control by spraying of Metasystox, wettable sulphur or kelthan. Thrips attack the flower which ultimately controlled by spraying of nuvacron or soil application of thime damage shoots. Tuberose is attacked by fungi masses on leaf surface or near the so loose green colour due to rotting which extend the plant become weak and bear fewer or no flowering. This is controlled by spraying of nuvacron or soil application of thimet. Leaf spot is another disease caused by *Alternaria polyantha* in tuberose, spraying of 1 per cent Bordeaux mixture, 0.4 per cent zineb or 0.5 per cent ziram is recommended for controlling the disease. Tuberose is also attacked by rust and powdery mildew which are not serious. Spray of sulphur is recommended in such cases.

Harvesting and Yield

Tuberose is largely cultivated for production of flower spikes and loose flowers. For marketing of flower spikes, the tuberose is harvested by cutting the spikes from the base when quite a good number of flowers open on the spike. Picking of individual flowers is completed by 4.0 AM to 8.0 A.M. It has shelf life of 2-5 days. The yield of fresh loose flowers per year per hectare for production of concrete in single tuberose varies from 12 to 15 tonnes. In the first two years, flower yield is high, and declines considerably in the third year is started.

For extraction of concrete and absolute, it requires solvent extraction process. The first extraction of fresh flower yields the floral concretes 0.2 and 0.12 per cent absolute. The absolute is a highly viscous or semi-liquid brownish mass with powerful and lasting odour, truly reminiscent and characteristic of fresh flowers. It may be mentioned here that steam or hydro-distillation of tuberose flowers can be carried out but produces only a very less yield of essential oil and very poor odour. The total yield of concerete from one ha. of land is 7 to 10 kg. Concrete of tuberose is a light to dark brown, waxy, quite hard mass, only partly soluble in high grade alcohol. Tuberose isolates contain methyl benzoate, methyl anthranilate, benzyl alcohol, benzyl benzoate, butyric acid, probably phenyl acetic acid, methyl salicylate, eugenol, geraniol, nerol both free and as acetates and farnesol.

Uses of Tuberose

The odor is intense, heavy sweet floral, honey with orange, green, oily-fatty undertone. Oil is produced rarely, normally absolute (by solvent extraction and further extraction by alcohol) is manufactured at industrial scale and is used.

It is used in fine fragrances for men and women both. The use level is 0.1 to 1.0%, very rarely perfumers use more than 1% in any formulation being costly material. Normally for sweet, heavy floral fragrances it is used.

It is allowed to be used in flavours but used only in traces.

The tuberose absolute is used in high quantity when cost allows to be used in high class floral fragrances of heavy and sweet type such as frangipani, lilac, heliotrope, gardenia, violet and heavy oriental types, oppoponax, in fantasy perfumes, etc.

The tuberose does not have much use in Aromatherapy.

ECONOMICS OF CULTIVATION OF TUBEROSE

Particulars per hectare	**Cost of Cultivation 1st year (Rs.)**	**Cost of Cultivation 2nd year (Rs.)**
A. Cost of Cultivation and Distillation		
Cost of seeds/plant material **(1,00,000 Bulbs)**	50,000	----
Land preparation for planting of seeds	4,000	----
Fertilizers	2,500	2,500
Labour cost	4,000	5,000
Irrigation (6)	4,000	5,000
Plant protection	500	500
Weeding (2 times manually)	2,000	2,000
Harvesting/flower picking	5,000	8,000
Transportation of flower	2,000	4,000
Distillation charges .	20,000	20,000
Miscellaneous	1,000	500
TOTAL	95,000	51,500
B. Gross Returns		
Concrete Yield (Kg)	7 kg/ha.	7 kg/ha.
Price of concrete (Per kg)*	Rs. 15,000/kg	15,000/kg
Cost of oil produced (per ha.)	1,05,000/-	1,05,000/-
C. Net Returns	10,000/ha./yr	53,500/ha./yr.

The above cost parameters are based on prevailing market rate of manpower, other materials and produce (Concrete) may deviate in future.

VETIVER

(*Vetiveria zizanioides*)

Cultivation Technology of Vetiver

Khus or Vetiver botanically known as *Vetiveria zizanioides L.* Nash is a perennial aromatic plant belongs to family Gramineae or Poaceae. It is known for highly valued essential oil obtained from their roots. The Vetiver plant is native to India, Pakistan, Bangladesh, Ceylon and Malaysia. In India, two types of essential oils produced i.e. North Indian type and South Indian type. North Indian oil has better perfumery value than south Indian type due the odour value and chemical constituents. The plant is about 2 to 2.5 meter tall, grows along the road side fellow spaces and railway tracks which provides possibilities for utilization of waste lands. It may survives in water logging as well as drought conditions (Plate 15).

The leaves blades are linear, 35-85cm. long, 5-25 mm wide, glabrous and pale green in colour. The highly scented aromatic oil is obtained through steam distillation as well as by hydro distillation (traditional*Deg-Bhapka* method) of 12-18 months old roots.

Soil and Climate

Vetiver plant grows in various types of soil, a fertile and well drained sandy loam soil is best suited for its cultivation. Water logged soils, with pH up to 9.0 and light soils containing 75-80% sand can be best utilized for economic production of vetiver crops. The root growth and development is more in medium textured soil, lengthy and fibrous roots can be obtained from light textured and sandy soil and the roots can be easily removed from soil layers during digging time. Vetiver grows best in hot and humid climate. In Tropical climate of south India, vegetative plant growth is better than north India as temperature during winter months (November- February) is low in latter and plant remains dominant. For optimum growth, plant requires long day conditions and plenty of sunlight. The shade adversely affects the plant growth and root development of crop.

Propagation

The Vetiver propagated both by seeds and vegetative parts by splitting the clumps into slips. However, it is desirable to plant the crop for commercial production by the vegetative means i.e. slips. For research purpose, seed propagation is generally meant for development of new varieties in crop improvement programme. The slips are prepared from 6 to12 months old clumps are cut from 25-35 cm. from the ground level. These clumps are then dug out and divided into slips containing 2-3 tillers. Slips are kept moist and stored in shade. Dried leaves are removed from the slips to avoid the mortality.

Planting

The plantation of vetiver can be done anytime of the year, except severe winter months (November-Mid January) because the temperature during these months remains very low and hot summer (May-June). The crop planted during rainy season does not require any irrigation, except just after planting. The optimum root growth and oil production, it is recommended to maintain a plant population of 44,000-45,000 plants per hectare with the spacing of 50cm line to line and 45 cm plant to plant distance. Since, it is a fast growing plant it does not allow any weeds to grow after 2-3 months from planting. One or two weeding is requires during the initial stages for its establishment. In good soils, planting is done at 50 cm. row to row and plant to plant 45 cm whereas in poor or waste soils planting is done at 45 cm row to row and 30 cm. plant to plant is recommended.

Fertilizer Application

Vetiver is wild grown plant so the fertilization is not required but for optimum root growth and oil production, fertilizer application is recommended depending upon the type of soil. In light textured soil, soils with high pH and laterite soils of south India application of 40kg. of N, 35 kg P_2O_5 and 25 K_2O/ha should be made before planting. During second year of crop growth, application of 35kg/ha should be applied after the rains in June-July.

Intercropping

In vetiver, intercropping during *kharif* season is found possible because it is slow in growth during initial 60 to 80 days period after planting. Intercropping may be done with leguminous crop namely cowpea (fodder crop) in vetiver to improves the root and oil yield and also economically benefited. The pigeon pea has been found to be the most economically viable intercrop in the vetiver field. On one hand where *kharif* crops show no adverse effect on the crop, *rabi* crops like mustard, chickpea and wheat have shown to reduce the oil yield and root growth.

Irrigation

Regular irrigation is not required in case of crop planted during the rainy season, except at the time of plantations. However, rains free period, irrigation supports is better for plant growth and root yield. Before digging of roots, when field is too dry, a light irrigation minimizes the cost of digging, by increasing the efficiency of labourers and more rootlets can be taken out and separated from the soil easily.

Harvesting/Digging of roots

After introduction of new varieties, the time of harvesting is 15-18 months after planting. Low soil moisture at the time of digging helps in separation of roots as soil can be shaken off from the roots easily. The digging of roots before 18 months gives tender and thin roots which yield low quality oil. The odour of such oils is earthy and green and is considered poor in quality. It is advisable to harvest the roots during the dry winter months to get the highest oil yield as well as best quality oil. Fully developed roots yield oil with high specific gravity and optical rotation, brown colour that gives richer and long lasting odour. The quality of roots and the oil is highly influenced by various factors such as fertility of soil, climatic conditions, quality and age of the roots, harvesting period, and the type of distillation technology.

Distillation

The extraction of essential oil from the roots is done by steam distillation and or hydro distillation. Freshly dug roots give better oil recovery but poor in quality, whereas semi-dried roots yield essential oil with better odour. Before distillation, roots are cut into small pieces of 5-6 inches in size & sprinkled with water. In India, three methods are common for distillation of Vetiver roots:

i) **Hydro-distillation or *Deg and Bhapka* method:** This method is adopted by traditional oil distillers of Kannauj. Depending upon the size of deg, 50-100kg., semi dried roots are distilled. The recovery of oil in this method is lower than the other methods (0.20- 0.25%) and the process is also time taking. But the quality of oil is superior known to perfumer.

ii) **Steam Distillation:** In this method, the steam is generated through boiler which is used for the distillation of roots. Boiler operated plants take 10-12 hrs to complete distillation process. In this method, the cost of production is higher than other methods.

iii) **Hydro-steam Distillation:** It is done by directly fired distillation units or Field Distillation Unit are easy to operate and the operational cost is also cheep because in this method several kinds of agro- waste including the distillation waste are used as fuel. It is recommended in these methods, the water coming out from separator can be cohobated for higher oil recovery.

Pests diseases and their control

The vetiver plant is highly tolerance to diseases and pests. However, a number of diseases have been reported on Vetiver. Leaf blight is appeared during rainy season. In dry areas plantation is also affected by the termites during the early stages of growth. Grubs of beetle have also been found infesting the roots. Spray of suitable pesticide like chloropyriphos with irrigation water is recommended to check the damage by these insects.

Yield and Oil Content

The yield of essential oil depends on root yield that depends upon the soil and duration of crop. Crop cultivated on normal soils gives 30-35q semidry roots per ha. from 18 month old plantation. Sandy and sandy loam soil yields about 22-28 q. roots/ha whereas from alkali soil with pH 9.0- 9.2, 13-16q. roots per ha are obtained. Oil content varies according to location and type of roots. Roots collected from wild sources in north India yield 0.15- 0.35%, whereas in south India it is 0.55-0.65%. The Cultivated roots of improved varieties yield 0.75- 0.85% oil. High yielding varieties may yield 12 to 24 kg/ha, depending upon the soil and management practices.

Uses of Vetiver Oil

The oil has two varieties north Indian and South Indian type. The odor is sweet, woody, earthy rich-heavy in both the varieties. The North Indian one is more earthy and fresh as compared to South Indian one.

It is used in fragrance for imparting wood-earthy note and also for fixation where color or misability with alcohol is not issue. Can be used upto 20% in fragrance, though general use level is upto 10%. Even in North Indian variety the oil do comes from two modes one by hydrodistillation (digand bhabka) and other by steam distillation. The oil by hydro distillation is more sweet and mellow in odor and used in fine fragrance. The oil out of steam distillation is also used in fragrances particularly in chypre and Fougere types. While South Indian is used in fragrances for soaps and cosmetics.

The vetiver oil and vetiverryl acetate are taken out of vetiver oil by fractionation. Both of them are very important ingredients for fragrances and flavours.

In aromatherapy the vetiver oil is also very important having antiseptic, antispasmodic, sedative and stimulant action. Quite useful for skin care particularly acne, cuts, oily skin. Effective in arthritis, muscular ache and pain, rheumatic problems and in sprain. Also known as oil of tranquility, used in depression and nervous tension.

In fragrances vetiver oil is used not only as fixative but as odor contribution to fragrance of fougere, chypre, woody, aldehydic or amber-aldehydic type. Also good for oriental bàses, woody notes, rose base. It blends well with oakmoss, lavender, clay sage, memosa, cassie and also with synthetics like ionones, iso eugenol, amy salicylate, linalool, cinnamic alcohol, etc.

Economics of Cultivation of Vetiver per Hectare

Particulars per hectare	**Cost of Cultivation (Rs./ha.)**
A. Cost of Cultivation and Distillation	
Cost of plant material **(44000 slips)**	22,000
Land preparation for planting	2,000
Fertilizers	2,000
Labour cost	2,000
Irrigation	2,000
Plant protection	500
Weeding (2 times manually)	2,000
Harvesting/Digging of roots	15,000
Transportation of roots	1,000
Distillation charges	7,500
Miscellaneous	1,500
TOTAL	57,500
B. Gross Returns	
Oil Yield (Kg)	12 kg/ha.
Oil Price (Per kg)	Rs. 10,000/kg
Total cost of oil produced (per ha.)	1,20,000
C. Net Returns	**62,500/ha./year**

The above cost parameters are based on prevailing market rate of manpower, other materials and produce (Essential oils) may deviate in future.

Chapter **4**

Distillation Technology of Aromatic Plants

Distillation defined as 'the separation of the components of a mixture of two or more liquids by the virtue of the difference in their vapour pressure'. It is process by which volatile odoriferous substances along with vapour is received in condenser where condensation process take place by flowing of cold water. The condensate is received in receiver/separator in a liquid from which is separated. For extraction of essential oil from plant material (root, flowers, whole plant, leaves, bark, etc.) different methods are followed. Most of the essential oils have been obtained by steam distillation, hydro-steam distillation or by Hydro distillation.

Hydro distillation (Deg-Bhapka Method)

It is a traditional method type of distillation technology, but it is still being practiced by the growers in Kannauj, Hasayan and Etah in Uttar Pradesh and Pushkar in Rajasthan for extraction of oil from rose, jasmine, henna, marigold, etc. In this process, the plant material/flowers are submerged in boiling water in the tank (*Deg*). The distillation apparatus is divided into three main parts called as deg (still), *bhapka* (receiver) and chonga (condenser-cum-connecting pipe). The still and receiver is made-up of copper which are connected by a bamboo which serve the purpose of condensation. The bamboo is wrapped by gunny bag to minimize the loss of heat through evaporation. This whole system is fixed on a furnace and directly heated by wood for distillation (Plate 16).

Hydro-steam distillation (Field Distillation Unit)

As the name, Field distillation, it is used for bulk oil of aromatic crops like basil, mints, palmarosa, lemongrass, citronella, *Tagetes minuta*, etc. In this method, distillation unit is divided into three main parts i.e still or tank made up by mild steel or stainless still, condenser and receiver-cum-separator are made up by stainless steel. The still is divided into two parts i.e false bottom perforated grid and actual bottom. The perforated grid is settled on 1.6 to 1.8 feet or more height from the actual bottom, the space between actual and false bottom is filled by water. The material which is to be distilled loaded on the perforated grid for distillation. During the operation of unit water level should be maintained at a specified level after every batch of distillation (Plate 16 & 17).

Steam distillation (Boiler Operated Unit)

In this process, essential oil is obtained by passing steam through material which is to be distilled. In this process, the steam is generated by a boiler, which is connected by a pipe to the still or tank. The pressure of steam is controlled according to the nature of plant material which is being distilled. The tank/still, condenser and receiver-cum-separator are made-up of stainless less steam. This process is comparatively costlier than hydro and hydro-steam distillation. Basically, this process is used for hardy and costly plant materials, such as sandal wood, nagarmotha, vetiver, kapoor kachri, jatamami.

Chapter **5**

Research & Development Institutions Working on Aromatic Plants

1. Central Institute of Medicinal & Aromatic Plants
   P.O. CIMAP, Near Kukrail Picnic Spot
   Lucknow-226 016 (U.P).
   Tel. 0522-2359623; 2359625
   Fax- 0522-2359666
   E-mail.: director@cimap.res.in
   Web. : www.cimap.res.in
2. Regional Research Laboratory,
   Industrial Estate,
   Trivendrum-695019
   www.rrltrivendrum.res.in
3. Indian Institute of Integrative Medicine (IIIM)
   (RRL-Jammu)
   Canal Road,
   Jammu Taawi,
   Jammu-180 001 (J&K)
   www.iiim.res.in
4. Regional Research Laboratory,
   Jorhat,
   Assam-785006
   www.rrljorhat.res.in

5. Regional Research Laboratory,
   Bhubaneshwar-751013 (Orissa)
   www.rrlbhuwaneshwar.res.in
6. National Chemical Laboratory
   Pune-411 008
   www.ncl.res.in
7. Institute of Industrial Toxicology Research
   Lucknow - 226010
   www.iitr.res.in
8. Indian Institute of Horticulture Research,
   Hessarghatta, Bangalore-560 089.
   www.iihr.icar.in
9. National Bureau of Plant genetic Resources,
   Pusa Campus,
   New Delhi-110 012.
   www.nbpgr.icar.in
10. Indian Agricultural Research Institute,
    Pusa Campus,
    New Delhi-110 012.
    www.iari.res.in
11. Kerala Agricultural University,
    Aromatic and Medicinal Plant Research Station,
    Odakkali, P.O Assamannur,
    Ernakulam,
    Kerla-683549.
12. Punjab Agricultural University,
    Ludhiana, Punjab-141004.
    www.pau.edu.in
13. Haryana Agricultural University,
    Hissar-125004.
    www.hau.edu.in

14. G.N.K.V.V. Collage of Agriculture,
Indore-452001.
15. University of Agriculture Sciences,
G.K.V.K, Bangalore-560065.
16. G.B. Pant University of Agricultural and Technology,
Pant Nagar, Nainital (U.P).
17. National Research Centre for Spice,
Calicut-673012.
18. Himachal Pradesh Krishi Vishva Vidhyalaya,
Solan-713212 (H.P)
19. Centre Food Technological Research Institute,
Mysore,
Karnataka.
www.cftri.res.in
20. Forest Research Institute,
Dehradun-248006 (U.P).
www.fri.edu.in
21. International Institute of Ayurveda,
Coimbatore, Tamil Nadu.
22. National Botanical research Institute,
Rana Pratap Marg,
Lucknow.
Tel. 0522-2205831-35; Fax: 0522-2205848
www.nbri-lko.org
E-Mail. director@nbri.res.in
23. Fragrance & Flavour Development Centre,
Industrial Estate, Makarand Nagar,
Kannauj - 209 726. (U.P)
Tel. 05694-234791/234465; Fax: 05694-235242
E-mail: ffdcknj@sancharnet.in
Web. www.ffdcindia.org

Chapter 6

Financing Agecies

Name of Financing Agencies who provide Subsidy on Cultivation and Processing of Aromatic Plants

1. **National Horticulture Board**

   National Horticulture Board
   (Ministry of Agriculture, Govt. of India)
   85, Institutional Area, Sector-18
   Gurgaon- 122 015
   Tel. 0124-2341866

2. National Medicinal Plant Board

   36, Chandralok Building
   (Ministry of Health and Family Welfare)
   Department of AYUSH)
   Janpath, New Delhi **- 110 001**
   Tel. 011- 23319255/23319360

SCHEMES OF NATIONAL HORTICULTURE BOARD

1. Objectives

- Review the present situation of horticulture development in particular area/ State.
- Development of primary/secondary data of various aspects on horticulture.
- Identify constraints and suggest their remedial measures.
- Development of short term and long term strategies for systematic development of horticulture.
- Provide consultancy services, expert services & establishing labs etc. in pursuance thereof.

2. Pattern of Assistance

100% cost of the study is to borne by the Board.

3. Nodal Organizations

- State Governments/UT's
- Central Government Organizations/agencies including NHB
- Other organizations through concerned State Government/UT's.

4. Aspects of Studies/Surveys

1. Techno Economic Feasibility Studies for development of horticulture in States/UT's/Belts/Zones etc.
2. Market Studies/ special problems/Area/Commodity based study/ other aspects.
3. Expert Services for project identification, formulation, implementation, monitoring & evaluation etc.
4. Multi-disciplinary and specific studies by NHB.
5. Emplacement of Consultancy Firms.

Study would be conducted through empanelled consultants. The penal of consultancy firms would be updated after every three years.

A) Criteria for empanelment of consultancy firms:-

1. Consultants should have capability, as may be evident from their past experience in similar work in agriculture/ horticulture/post-harvest management.
2. The consultants must have required infrastructure for taking up consultancy as per job requirements.
3. The consultants must have expertise and experience in the field of horticulture, financial management, project formulation, project evaluation, monitoring etc.
4. Preference would be given to the such consultancy firms, who have experience in conducting studies in a similar agro-climatic conditions.

5. The consultancy firms should be registered, and it should be multi disciplinary, having minimum experience of 3-5 years.
6. Individual consultants or subject matter specialist like horticulture/agriculture experts, economist/ management/ legal/personnel/ financial/ marketing/ information-technology experts can be engaged for specific activities, where the expertise of an individual is required.
7. The selection of experts for specific/multi-disciplinary studies shall be finalised by the committee of NHB keeping in mind their known professionalism in the field.

B) Procedure for Empanelment of consultancy firms

- Wide publicity would be given through mass media for calling the offers from consultancy firms for empanelment.
- Based on criteria, a panel of consultancy firms would be prepared by Empowered Committee of the Board and would be got approved by Managing Committee excepting for the experts mentioned at Sr. No. 7 above.
- Multi-disciplinary and specific studies would also be entrusted to registered bodies having a minimum experience of 3 to 5 years and Individual experts in the field of horticulture/agriculture/economics/finance/legal/personnel/ Marekting/information Technology etc. by NHB for specific activities, where the expertise is required as mentioned at Sr. No. 6 & 7 above.

6. Procedure for award of studies/surveys

- Consultants would send their detailed offer taking into consideration scope of work/term of reference within prescribed period.
- Consultants would be required to give offer in-respect of Technical & financial aspects.
- Offers would be scrutinized by the Board for consideration of Empowered committee.
- Consultants would present their case in person before the Empowered Committee on the date & time to be decided by the Board.

- The study would be carried out in accordance with the scope of work.
- Period of study would be decided by the Empowered Committee on case to case basis.
- Study would be warded by the Board on the recommendation of the Empowered Committee.

Above mentioned procedures shall not apply in the case of selection of experts by the internal committee of NHB.

7. Mode of Payment

- 25% on award of work.
- 50% on submission of Draft Report.
- 25% on submission of final report after acceptance by Board/concerned State Government/Organization

However, the consultant may furnish an indemnity bond and personal guarantee before the release of payments.

NOTE:- For any delay in submission of Draft Report, penalty @ 1% of total fee may be levied on the consultants for every delay of one month or part thereof.

8. Submission of Draft Report

- Draft report would be submitted to the consultants within the prescribed time.
- Consultants would also present the study undertaken through Slides/- etc. before the NHB & concerned sponsoring organization.
- Comments of NHB/Sponsoring organization would be included in the study report.
- Consultants would coordinate for organizing the presentation of the reports.

9. Submission of Final Report

Consultants would require submittingtwo copy of Draft Report and 20 copies of Final Report duly bound alongwith computer floppy.

TECHNOLOGY DEVELOPMENT AND TRANSFER FOR PROMOTION OF HORTICULTURE

Eligible Components

1. Objectives

- Popularization of new technologies/tools/techniques for commercialization/ adoption.
- Introduction of new concepts to improve farming systems.
- Upgradation of skills by exchange of technical know-how.
- Consolidating research efforts for specific problems.
- Identification/collection, rapid multiplication & popularization of indigenous crops and other flora of horticultural importance with emphasis on domestic and export promotion.
- Familiarization and exposure towards the newer scientific concepts/temper and research & development from hitherto unknown, unexplored and traditional status of farming and post harvest management on to the modern lines.

2. Eligible Components

1. Introduction of New Technologies

A. Pilot Project based on high quality commercial production, compact area development approach for popularizing new and modern scientific concepts in horticulture.

i. Commercial sustenance shall be the main criteria for selecting such pilot projects.

ii. Projects should be smaller preferably below Rs. 10.00 lakhs.

iii. The project may have one of the following aspects:

  i. Development prototype for adoption on a commercial scale.

  ii. Demonstration and training on new technology/tool/ equipments.

  iii. Implementation of technology/concept.

iv. Technology/tool/equipment/critical input including planting material (imported/ indigenous) to be tested preferably on farmers' field for adoption & commercialization.

v. Funding requirements for each item to be indicated with justification.

B. Post Harvest Management, Primary Processing, Biotechnology and Introduction of New tool/equipment/machinery is to be considered on merit with one or all the following broad considerations:-

   i. Increase in efficiency of operation.
   ii. Contribution to quantity and quality.
   iii. Reduction in cost and contribution to economy.
   iv. Increasing shelf-life.
   v. Benefit such introduction over the existing level.
   vi. It should be relatively advanced tool/machinery/ equipment.

C. R & D projects on specific problems would also be considered on merit basis. Project based on indigenous fruit, vegetables, flowers, medicinal and aromatic plants would also be encouraged for domestic and export promotion.

NGO's are not eligible for R&D projects.

Pattern of Assistance

Grant upto 100% of the cost of project would be provided under the scheme. However, it shall be restricted upto:-

- Rs. 10.00 lakh in case of projects falling under Category (A) and
- Rs. 25.00 lakh for projects falling under Category (B) & (C) to State Agricultural/ Horticultural Universities/Research Organizations/ State Department of Horticulture/ Agriculture/PSU's and other related agencies/organizations for transfer and demonstration of proven technology on commercial scale and R&D efforts on specific issues.

However, staff component shall have to be supported by the concerned organization.

NHB may co-opt. the services of experts if required for sanctioning such projects.

2. Visit of Progressive Farmers

A) National/Domestic Visits:

i) Representative of State department of Agriculture/Horticulture may accompany farmers groups to be recommended by the State Govt. NHB's officers of concerned State/UT's may also organize such visits.

ii) Such groups should be actively involved in the development and promotion of horticulture.

iii) Farmers should be in the age group of 20 to 55 years.

iv) Farmers should have his own land where he/she may have been raising horticultural crops or intends to establish horticulture farm/horticulture crops/horticulture industry.

v) Preference should be given to small, marginal and women farmers.

Pattern of Assistance

Financial assistance in the form of grant-in-aid shall be provided for organizing such training/visit upto a period of 10 days (excluding journey period). Assistance shall be limited to 'to & fro' rail fare of second class sleeper/ordinary bus fare in addition to boarding charges @ Rs. 100/- per day per farmer upto a group of 30 farmers.

A committee of NHB shall finalize such proposals.

C. International Visit

NHB may coordinate the visit of progressive farmers to go abroad for exposure and training in modern horticulture related technology.

Pattern of Assistance

A group of 10 farmers can be taken with NHB assistance limited to economic class to & fro air fair.

Local hospitality to be borne by farmers themselves.

Only State Govt./NHB promoted proposals are to be considered.

A committee of NHB shall finalise such proposals.

3. Promotional and Extension Activities

- To demonstrate modern/scientific techniques/technologies at suitable locations/areas by the NHB.
- To demonstrate improved/high yielding varieties of fruits, vegetables, flowers, etc.
- A committee of NHB shall examine such proposals.

4. Export's Services from India/Abroad

i. Criteria for selection of experts would be based on technical development programme.

ii. Services of Experts would be permitted on individual project basis for a period not exceeding 15 days per year for first two years only.

iii. Selection of Experts be done by a Committee of NHB consisting of representative from NHB, ICAR & Horticulture Division, DAC.

Pattern of Assistance

Expenditure on expert's services for travel and per diem shall be in accordance with the Government of India's approved rates.

5. Technology awareness

- To create awareness about the introduction and new technologies/concepts by organising seminars/workshop/exhibition at the local/district level for promotion of horticulture.

- The amount sanctioned for a particular year is required to be spent within the same financial year giving proper justification, unless revalidated.
- NHB shall be given full opportunity to high-light NHB's activities during such seminars/workshops.
- NHB's name should be acknowledged properly in the publication of the material/prizes/shields distributed by the organisation.
- The organisation would supply five copies of the publication to the NHB, free of cost.

Pattern of Assistance

- Up to Rs. 50,000/- per events/ organisation to be considered by the Committee of NHB.

6. Organization/Participationin Seminars/Workshops/Exhibition, etc. for Promotion of Horticulture.
    - To create awareness about the introduction and new technologies/concepts by organising seminars/workshop/exhibition at the state, national and international levels for promotion of horticulture.
    - The amount sanctioned for a particular year is required to be spent within the same financial year giving proper justification, unless revalidated.
    - NHB shall be given full opportunity to highlight NHB's activities during such seminars/workshops.
    - NHB's name should be acknowledged properlyin the publication of the material/prizes/shields distributed by the organisation.

The organisation would supply five copies of the publication to the NHB, free of cost.

MARKET INFORMATION SERVICE FOR HORTICULTURE CROPS

Objectives

- The genera information on wholesales prices, arrivals and trends in various markets of the country for important fruits, vegetables & flowers, etc.

- To establish a nation-wide communication network for speedy collection and dissemination of market information data for its efficient and timely utilization.
- To develop a sound marketing strategy especially by making use of statistics generated by various Marketing Boards/ APMCs for optimizing returns to the producers.
- To analyze the trends of arrivals, prices and other related factors of the fruit and vegetable markets all over the country.
- To collect and disseminate information on international prices prevailing in potential foreign markets.
- To establish Web-sites and internet connection through efficient channels/systems for dissemination of national & international data.

Manner of Implementation

- At present, all the 33 Market Information Centres are collecting the information on prices & arrivals in their respective markets and communicating to the Central Coordinating Cell of NHB H.q. at Gurgaon which is compiling and analyzing the prevailing prices and arrivals of selected horticultural commodities of commercial importance.
- The information is received by E-mail and fax from the centers and is also disseminated back through the same means of communications on the very same day for further dissemination through mass media.
- Presently all the commercial and seasonal fruits & vegetables are covered for dissemination. Effort shall be made to cover more & more bulk varieties of fruits, vegetables and flowers of particular areas.
- Besides collecting the data on prices & arrivals of fruits and vegetables, Board is also forecasting the crop production for the coming year and compare the present data of different centers.
- A CD ROM namely Horti. Biz India shall be prepared in a professional way for the benefit of farmers, traders, ex-

porters, scientists and professional etc. which would include information about the high quality production, packaging, post-harvest management, cool chain, domestic and export markets etc.

- All valuable data collected at different NHB centers are published through monthly, weekly and daily bulletin regularly. It is proposed to publish horticulture data as a whole through publication of different database, commodity bulletin, leaflets, books, journals, newsletters etc.
- The data compiled and analyzed would be transmitted through Internet, NICNET or Web-sites or other systems established under this scheme.
- The international prices of fruits, vegetables and flowers are also collected from ITC Geneva and disseminated through publications.
- List of publication is at Annexure-8.

Important Fruits and Vegetables Covered

APPLICATION FORM FOR SEEKING FINANCIAL ASSISTANCE FROM NATIONAL HORTICULTURE BOARD UNDER SCHEMES "TECHNOLOGY DEVELOPMENT AND TRANSFER FOR PROMOTION OF HORTICULTURE AND ESTABLISHMENT OF NUTRITIONAL GARDEN IN RURAL AREAS

A. General Information about beneficiary organizations

1. Name the organization with full address
2. Status
3. Name of the person with whom correspondence is to be made, together with his telephone and telegraphic address/ Fax number, if any.
4. Year of establishment (wherever applicable)
5. Whether registered under the Registration of societies act, 1860 or any similar state statute (Registration number & date to be given) attach an certificate by gazetted officer in case of NGO's). [whether applicable]

6. Major objectives of organization. Attach certified copy of bye-laws of the society by Gazetted office in case of NGO's. [wherever applicable].
7. Total number of the members in the organization as on 31st march (year to be mentioned) in case of NGO's, association of growers [wherever applicable]
8. Summary of the activities undertaken by the organization in the last year.
9. Audited balance sheet of last 3 years duly certified by C.A. [wherever applicable].
10. Other sources from which financial assistance is received by the organization.
11. Whether the organization agrees to maintain a proper separate account of the grant sanctioned by the board.
12. Whether applied for assistance previously, if so, with what results.
13. Details of financial assistance already received from the board, if any, and the purpose for which it was received.

year	Amount of grant	Purpose in brief total exp. Incurred	total exp. Incurred	Amount of grant utilized	Has utilization certificate been accepted by the board	Remark

A. Project proposal
i. Details of the project proposal with justification for implementation
ii. Requirement of funds
iii. Total financial assistance sought from the board
iv. Has any grant of the same purpose been applied for from any other sources if so, please state

A. Supporting documents and annexures relating to project proposal

Certified that the information given above is correct and we agrees to abide by the rules and regulations of the board for utilization of the financial assistance sought for

Signature of Authorized Authority

References

1. Anonymous. 1961. Indian standard specification for oil of lemongrass (East Indian Oil of Lemongrass) (Revised) IS 327 : 1961. Indian Standards Institution, New Delhi-1.
2. Anonymous. 1961. The extraction of tuberose flo No. 47. pp. 1-3.
3. Anonymous, 1983. Studies on NPK requirements of lemongrass (Bangalore). Annual Report 1982-83. Central Institute of Medicinal and Aromatic Plants, Lucknow. pp. 107.
4. Anonymous, 1984. Regional Research Laboratory, Jammu. *Newsletter*, 1 (2) : 9.
5. Anonymous. 1985. Effect of different levels of nitrogen on yield and quality of lemongrass (Lucknow). Annual Report 1984-85. Central Institute of Medicinal and Aromatic Plants, Lucknow. pp. 118.
6. Anonymous. 1988. Annual Report 1987-88. Central Institute of Medicinal and Aromatic Plants, Lucknow. pp. 161.
7. Anonymous, 1988. IS 526. Indian Standard for oil of palmarosa. Bureau of Indian Standard, New Delhi.
8. Anonymous. 1989. Indian standard 528 for oil of *Mentha arvensis* (BSS, Delhi).
9. Anonymous. 1992a. Indian Standard 3134 for Menthol specification. pp. 1-3.
10. Anonymous. 1992. *Ocimum canum* Sims: RRL OC-11, RRL, Jammu, pp. 1-10. Atal, C.K. 1979. Essential oils in the1980's. *Indian Perfumer*, 23 (3) : 141-61.
11. Acosta, L. and Granda, M. 1982. Notes on cultivation of medicinal plants in Cuba. *Metricaria chamomilla* L. cultivos Tropicales, 4 (4) : 727-737.
12. Anup Kumar, Gauniyal A.K. and Virmani, O.P. 1986. Cultivation of *Pogostemon patchouli* for its oil. *CROMAP*, 8 (2) : 79-86.
13. Arak, E., Tammeorg, T. and Myaeorg. 1980. Dynamics of some components of chamomile essential oil. *Tarter Ulikooli Toimetised*, 523 : 19-32.
14. Akhila, A., Tyagi, B.R. and Naqvi, A. 1984. Variation of essential

oil constituents in *Cymbopogon martini Wats.* (var. *motia)* at different stages of plant growth. *Indian Perfum.,* 28 (3&4) : 126-28.

15. Angadi, S.P. 1990. Annual progress report of the project on mutation breading in scented geranium *(Pelargonium graveolens* L. Herit) for higher oil yield, quality and disease resistance, IIHR, Bangalore.
16. Armugam, R. and Kumar, N. 1979. Geranium cultivation in Kodaikanal hills. *Indian Perfumer,* 23 (2) : 12830.
17. Bailey, L.H. 1930. *Polianthes. The Standard Gycl* York. 3 pp. 2731·33.
18. Bailey, L.H. and Bailey, E.Z. 1977. *Polianthes. H* 893.
19. Baker, J .G. 1888. *Handbook of the Amaryllideae Ir* London.
20. Balyan, S.S. and Pushpangadan, P. 1988. A study on the taxonomic status and geographic distribution of the genus *Ocimum.* PAFAI, 10 (2) : 13-19.
21. Barua, A. and Bordoloi, D.N. 1983. Record of a new disease of lemongrass *(C.flexuosus)* caused by *Curvularia verruciformis. Curr. Sci.,* 52: 640-41.
22. Banker, G.J. and Mukhopadhyay, A. 1985. Response of *Polianthes tuberosa* L. cv. Single to high doses of NPK. *South Indian Horticulture,* 33 (3) : 214-16.
23. Banker, G.J. and Mukhopadhyay, A 1990. Effect of NPK on growth and flowering in tuberose cv. Double. *Indian J. Hortic.,* 47 : 120-26.
24. Bentham, G. and Hooker, J.D. 1873. *Genera - Plantarum,* Vol. 11. *Reeve* ani Co., London.
25. Biswas, K. 1967. *Indian Farming,* October, 19-22.
26. Bhattacharjee, S.K., Mallik, R., Yadav, L.P. and Mukherjee, T. 1981a. Response of *Polianthes tuberosa* Linn to nitrogen, phosphorus and potash fertilization. *Bull Botan. Soc. Bengal,* 35 : 75-80.
27. Bhattacharjee, S.K. and Mukherjee, T. 1979. Effect of some growth regulators on *Polianthes tuberosa* Linn. *Lal Bough J.,* 24 (4) : 30-35.
28. Bhattacharjee, S.K., Mukherjee, T. and Yadav, L.P. 1981b. Testing of *Polianthus tuberosa* Linn. cultivars for cut flowers. *Lal Bough J.,* 26 (2) : 52-53.
29. Bhattacharjee, S.K., Mukhopadhyay, T.P. and Yadav, L.P. 1979a. Effect of nitrogen, phosphorus and potassium on growth and

flowering of Amaryllis and Tuberose. Sand culture. *Lal Bough J.* 24 (4) : 36-43.

30. Bhattacharjee, S.K., Yadav, L.P. and Mukhopadhyay, T. 1979: Effect of bulb size, planting depth and spacing on tuberose *(Polianthes tuberosa* Linn). *Lal Bough J.,* 24 (4) : 24-27.
31. Bhujbal, B.G. and Whaval. 1978. Effect of number of bulbs per hill on flower production in tuberoses *(Polianthes tuberosa* L.) *Indian J. Om. Hort.,* 2 : 27-28.
32. Bose, T.K., Jana, B.K. and Moulik, S. 1987. A note of 'the micro-propagation of tuberose from scale stem section. *Indian J. Hortic.,* 44 (1 +2) : 100-01.
33. Brundell, D.J. and Steenstra, D.R. 1986. The effects of protected cultivation, pre-sprouting and lateral tuber removal an tuberose production. *Acta Horticulturae,* 177 : 361-67.
34. Bhattacharjee, S.K. 1982 *Annual Report,* Indian Institute of Horticultural Research, Bangalore pp. 86.
35. Bhattacharjee, S.K. 1983. Standardization of agro-technique in jasmine. *Annual Report,* Indian Institute of Horticultural Research, Bangalore. p. 86.
36. Bhattacharjee, S.K. 1984. Standardization of agro-technique of jasmine. *Annual Report,* Indian Institute of Horticultural Research, Bangalore. pp. 73.
37. Bhattacharjee, S.K., 1985. Effect of plant densities on flower yield of *J. grandiflorum. Indian Perfumer,* 29 (3-4) : 177-80.
38. Bhattacharjee, SK and Balakrishna, M. 1983. Standardization of propagation of jasmine from stem cutting. II. Effect of rooting media, leaf number, length and woodiness of cuttings on root formation and survival of rooted cuttings. *Indian Perfumer,* 27 (1) : 36-42.
39. Bhattacharjee, S.K. and Balkrishnan, M. 1983. Standardization of propagation of jasmine from stem cutting. III. Multiplication of different species with the aid of growth regulators. *Indian Perfumer,* 27 (1) : 43-46.
40. Bhattacharjee, S.K. and Divakar, N.G. 1981. *Annual Report,* Indian Institute of Horticultural Research, Bangalore. pp. 105.
41. Bhattacharjee, S.K. and Divakar, N.G. 1983. Flower yield and essential oil content of *J. grandiflorum* L. as influenced by nitrogen,

phosphorous and potash fertilization. *IX International Congress of Essential Oils.* Singapore. Book-1. pp. 121-25.

42. Bhattacharjee, S.K., Sharma C.K. and Divakar, N.G. 1979. *Annual Report,* Indian Institute of Horticultural Research, Bangalore. pp. 137-38.
43. Bhupal Rao, J.V.R. and Divakar, N.G. 1975. *Annual Report,* Indian Institute of Horticultural Research, Bangalore. pp. 54.
44. Bhupal Rao, J. V. R. and Divakar, N. G. 1977. *Annual Report,* Indian Institute of Horticultural Research, Bangalore pp.73.
45. Bhupal Rao, J.V.R. and Divakar, N.G. 1979. A note on seed characteristics of different *Jasminum* species. *South Indian Horticulture.*
46. Behura, S., Sahoo, S. and Dutta, P.K. 1988. Effect of organic manure for growth and oil Yield in chromita over burdens. *Indian Perfum.,* 32 (2) : 153-57.
47. Blajek, Z. and Stary, F. 1961. Some data on the composition of flowers of the pure chamomile from C. zechoslovakia. *Pharmazie,* 16 : 477-82.
48. Chandra, V., Mishra, P.N. and Kapoor, L.D. 1970. Blue oil from *Matricaria chamomilla* L. A new industrial crop. *Indian Perfumer,* 14 (1) : 27-31.
49. Chandra, V., Mishra, P.t-.: and Singh, A. 1979. *Matricaria* blue oil. Extension Bulletin No. 4. Economic Botany Information Service. National Botanical Research Institute, Lucknow.
50. Chandra, V., Singh, A. and Kapoor, L.D. 1983.'Experimental cultivation of some essential oil bearing plants in saline soils. *Perfum. Essential. Oil Rec.,* 59 : 871.
51. Chadha K.L. and Gupta R. Advances in Horticulture, (Medicinal and Aromatic Plants-Vol.-11(1995).
52. Chopra, R.N., Nayar, S.L. and Chopra, I.C. 1956. Glossary of Indian Medicinal Plants. C.S.I.R., New Delhi.
53. Chandravadana. M.V. and Vasantha Kumar, T. 1989. The yield and oil quality of scented geranium as affected by biomass storage duration and different temperatures. *Indian Perfumer,* 33 (4) : 246-48.
54. Chandra, M.K., Singh, J.N. and Nand, Kewala. 1983. Effect of nitrogen levels and harvesting time on the growth and yield of

Japanese mint *(Menha arvensis Linn.). Indian Perfum.*, 27 (2) : 94-98.

55. Chauhan, H.S., Ram, P., Singh, U.V. and Singh, K. 1991. Response of Japanese mint *(Mentha arvensis* L.) to phosphorus and zinc on mollisoles of *Tarai* belt in Uttar Pradesh. *Indian Pefum.* 35 (3) : 15055.
56. Chandthoke, N. and Ghatak, B.J.R. 1969. Studies on *Tagetes minuta* : Some pharmacological action of the essential oil. *Indian J. Med. Res.*, 57 (5): 864-76.
57. Chopra, I.C., Nigam, M.C., Kapoor, L.D. and Handa, K.1. 1963. Indian Tagetes oils. *Soap Perfumers Cosmetics,* 36 : 686-89.
58. Chandra, V and Srivastava, H.P. 1977. Package of practices for cultivation of *Jasminum granqiflorum* L. *Indian Perfumer,* 21 (2) : 91-93.
59. CIMAP Newsletter. 1987. A high yielding tetraploid strain of vetiver developed. 14 (2) : 2.
60. Chinnamma, N.P. and Aiyer, R.5. 1988. Effect of fertilizers and harvest intervals on yield, nutrient composition and nutrient uptake of palmarosa. *Indian Perfum.*, 32 (4) : 270-87.
61. Chinnamma, N.P., Nair, E.V.G., Aiyer, R.S. and Saraswathy, P. 1988. Effect of fertilizers and harvest intervals on yield, nutrient composition and nutrient uptake of palmarosa. *Indian Perium.*, 32 (4) : 278-87.
62. Chandra, V. 1973. Cultivation of plants for perfumery industry at Lucknow. *Indian Perfum.*, 16 (1) : 40-44. Cuellar, A. and Gomez Gonzalez, C. 1973. Sapogenins from *Polianthes tuberosa* white lily variety. *Rev. Cubana Farm.*, 7 (2+3) : 125-33.
63. Das, AC. 1961. Diseases of Rajanigandha *(Polianthes tuberosa* L.) and Larkspur *(Delphinium ajacis* L.) caused by *Sclerotium rolfsii* Sacco *Sci. Cult.*, 27 (11) : 549-50.
64. Patra DD, Anwar A, Chand S, Chattopadhyay A, Prasad A, Pandey P, Kumar A, Singh A, Srivastava RK, Krishna A, Singh V, Tomar VKS, Bansal RP, Singh AK, Singh K, Bahl JR and Kumar S. Use of mint essential oil as an agrichemical: Control of N-loss in crop fields by using mint essential oil – coated urea as fertilizer. *Current Science.* (2001): 81 (12): 1526-1528.
65. Desai, B.L. 1957. Flowers that fill your garden with fragrance. *Indian Farming,* 7 (1) : 8-11.

66. Dhingra, D.R., Gupta, G.N. and Shukla, U.N. 1956. Indian Jasmine oils. *Indian Soap Tour.*, 21 : 219.

67. Durvelle, I.P. 1923. *The Preparation of Perfumes and Cosmetics.* Scott., Greenwood and Sons, London. pp. 69-70.

68. Doreswamy, K. and Sundaram, M. 1978. A new disease resistant variety of geranium. *Indian Perfumer,* 22 (1) : 43-45.

69. Doreswamy, K. Sundaram, M. and Maniam, K. 1979. Root knot nematode control in scented *Pelargonium graveolens* (L.) Hir. *Madras Agric J.*, 66 (3) : 205-6.

70. Dureswamy, P. and Armugam, R. 1980. Preliminary studies on the rooting of geranium cuttings as affected by soil reaction. *Indian Perfumer,* 24 (3) : 168-69.

71. Downum, K.R. and Towers, G.H.N. 1983. Analysis of thiophenes in the Tagetes (Asteraceae) by HPLC. *J. Natu. Prod.*, 48 : 98-103.

72. Dutta, P.K. and Singh, A. 1964. Effect of different spacing on fresh flower and oil yield of *Matricaria chamomilla. Indian J. Agron.*, 9 (1) : 11-12.

73. Everett, T.H. 1975. *Polianthes. The New Illustrated Encyclopedia of Gardening.* Greystore Press, New York. Vol. 16. pp. 2430-31.

74. French, M.G. 1952. Plants from the whole world contribute to the making of perfume. *Gdn J.*, 2 (2) : 4750.

75. Franz, C. 1981. Content and composition of the essential oil in flower heads of *Matricaria chamomilla* L. during its ontogenetical development. Technische Universitute Muuchen D-8050, *Freising welhoustephan* (GFR) 317-321.

76. Franz, C. and Holzl, J. 1978. Variation in the essential oil of *Matricaria chamomilla* L. depending on plant age and stage of development. *Acta Hortic,* **73** : 209-38.

77. Franz, C., Holzl, J. and Kirsch, C. 1983. Influence of NPK fertilization of *Chamomilla. (Chamomilla recutita.* syn. *Matricuria chamomilla* L.) II. Effect on the essential oil. *Gastenbauwissenschaft.* (GFR), **48** (1) : 17-22.

78. Franz, C. and Kirsch, C. 1974. Growth and flower formation by *Matricaria chamomilla* in relation to varied nitrogen and potassium nutrition. *Gartenbauwissenschaft* (GFR), **39** (1) : 9-19.

79. Ghauri, E.G. 1984. Correct scientific name of 'Babuna' used widely as a drug in Unani system of medicine. *Bangladesh J. Sci.*

Industr. Res., 19 (1-4) : 193-200.

80. Guenther, E. 1949. *The essential Oils,* Vol. 1. Van Nostrand Co., Inc., New York.
81. Guenther, E. 1972. *The Essential Oils,* vol. V. D Van Nostrand Company. Inc., New York. pp. 438-45.
82. Gulati, B.C. and Duhan, S.P.S. 1982. Cultivation of *Pelargonium graveolens* as annual crop. In *Cultivation and Utilisation of Aromatic Plants* (Atal, C.K. and Kapur, B.M., Eds.). RRL Jammu-Tawi. pp. 55965.
83. Gulati, B.C., Duhan, S.P., Gupta, R. and Bhattacharya, A.K. 1977. Introduction of French basil *(Ocimum basilicum* L.) in Tarai of Nainital, U.P. (India). *Perfumeric and Kosmetics,* 58 : 165-69.
84. Gupta, R. 196.5. Japanese mint-A new industrial crop in India. *Indian Perfum.*, 9 (pt I) : 69-75.
85. Gupta, A. and Agarwal, M.K. 1963. *Odontotermis obesus* Ramb. as a pest of Japanese mint. *J. Bombay Nat. Hist. Soc.*, 60 (1) : 285-87.
86. Gupta, R. 1976. Mentha brings money to Indian farmers. *Financing Agril.*, 7 (4) : 21-24.
87. Gupta. R., Gulati, B.C. and Singh, J.P. 1970. Mentha hold great promise. *Indian Farmers Digest.*, 3 (9) : 43-45 and 49.
88. Ganguly, D. and Pandotra, V.A. 1962. Important diseases of mints and their control. *Bull. Regional Res. Lab., Jammu.*, 1 (1) : 52764.
89. Gulati, B.C. and Duhan, S.P.S. 1971. Japanese mint in Nainital of Uttar Pradesh *Indian Perfum.*, 15 (pt 1) : 1-11; (pt 2-) : 15-30 and (pt 3) : 31-51.
90. Gulati, B.C. 1963. Present position of palmarosa and ginger grass oils industry: Scope of systematic cultivation of palmarosa. *Indian Oil and Soap J.*, 29 (4) : 87-101.
91. Gulati, B.C., Gupta, A and Duhan, S.P.S. 1970. Trial cultivation of palmarosa *(Cymbopogon martini var. mofia)* in Tarai and Bhaber tracts of UP. *Indian Oil and Soap J.*, 30 (10) : 105-14.
92. Gupta, A., Maheshwari, M.L., Singh, A.P., Mohan, J., Gupta, G.K. 1978. Effect of fertilizers on yield, oil content and oil composition of palmarosa as influenced by seasonal variations. *Indian Perfum.*, 22 (2) : 79-87.
93. Geetha Singh and Hippalgaonkar, K.V. 1992. Agro-technique for raising rooted branch cuttings of patchouli *(pogostemon cablin*

Benth.) for cultivation. *Indian Perfum.*, 36 (4) : 250-53.

94. Geetha, Singh and Hippalgaonkar, K.V. 1993. Influence of foliar applied kinetin as growth and essential oil content of patchoule *(pogostemon cablin* Benth.). *Indian Perfum.*, 37 (2) : 167-70.

95. Guenther, E. 1952. Recent developments in essential oil production-Patchouli oil. *Econ. Bot.*, 6 (4) : 355.

96. Gulati, B.C. 1963. Essential oil of Patchouli: Present position in India and scope of its development. *Indian Perfum.*, 7 (1) : 29-42.

97. Guenther, E. 1972. *The Essential Oils,* Vol. 4. Robert E. Krieger Publishing Co. Inc, Huntington, New York. pp. 156-80.

98. Gupta, R.S., Trivedi, K.C., Verma, S. and Gangrade, S.K. 1983. Studies on essential oil of vetiver *(Vetiveria zizanioides)* hybrid clones. *Indian Perfum*, 27 (2) : 104-07.

99. Gopalswamiengar, K.S. 1970. The tuberose. *Complete Gardening in India.* G. Kasturi Rangan, Bangalore. pp. 88.

100. Guenther, E. 1952. Concrete and absolute of tuberose. *The Essential Oils. D. Van Nostrand Co. Inc.*, New York. Vol. 5. pp. 343-48.

101. Guenther, E. 1956. *Chemical Abstract,* 50 : 8971.

102. Guenther, E., 1955. *The Essential Oils* Van Nostrand Co., New York. pp. 324.

103. Handa, K.L., Chopra, M.M. and Nigam, M.C. 1963. The essential oil of *Tagetes minuta* L. *Perfum. Essent. Oil Rec.*, 54 : 372-74.

104. Hussain, Akhtar, Virmani, O.P., Sharma, A., Kumar, A. and Mishra, L.N. 1988. *Major Essential Oils of India.* Central Institute of Medicinal and Aromatic Plants, Ram Sagar Mishra Nagar. Lucknow. India. pp. 135-40.

105. Hellyer, A.G.L. 1956. *Sanders' Encyclopedia of Gardening,* 22nd edn. W.H.L. Colling ridge Ltd. pp. 389. Hooker, J.D. 1960. *Index Kewensis.* Clarendon Press, Oxford, Vol. 2. pp. 584-85.

106. Hariprakasa Rao, M. and Vasantha Kumar, T. 1989. Effect of different levels of phosphorous on yield and P content in different patchouli cultivars. *Indian Perfum.*, 33 (1) : 8-13.

107. Hazarika, J.N., Barua, A. and Barua, A.KS. 1978. Effect of N,P,K fertilizers on the yield and quality of oil of palmarosa *(Cymbopogon martini* Stapf. var. *mofia)* under the influence of seasonal variations. *Indian Perfum.*, 22(1) : 36-39.

108. Hazarika, J.N. and Bora, A.C. 1977. Effect of varying combinations of N.P and K fertilizers on yield of palmarosa *(Cymbopogon martini* Stapf. var. *motia.). Indian Perfum.*, 21 (1) : 51-53.

109. Hazarika, J.N., Mishra, B.P. and Bora, A.C. 1981. Effect of spacing on yield of herb and oil of palmarosa *(Cymbopogon martini* var. *motia)* grown under Jorhat. *Indian Perfum.*, 25 (3&4) : 56-59.

110. Hamidii, A.E.L., Saleh, M. and Hamidii, H. 1965. The effect of fertilizer levels on the growth, yield oil production of *Matricaria chamomilla. Lalydea,* **23** (4) : 245-51.

111. Handa, K.L., Chopra, L.C. and Akrol, B.K. 1957. Introduction of some of the exotic aromatic plants in Jammu and Kashmir. *Indian Perfumer,* 1 : 42-49.

112. Holzl, J. and Demuth, G. 1975. The influence of ecological factors on the composition of the essential oil and the flavones in *Matricaria chamomilla* from different source. LA. critical composition of methods for quantitative determination. *Planta Medica,* **27** (1) : 37-45.

113. Hussain, A., Virmani, a.p., Sharma, A., Kumar, A. and Misra, L.N. 1988. Chamomile oil (German). In *Major Essential Oil Bearing Plants of India.* CIMAP, Lucknow. India. pp. 45-48.

114. Husain A., etal., Major essential oil Bearing plants of India (Central Institute of Medicinal & Aromatic Plants, P.O. Ram Sagar Mishra Nagar, Lucknow, 1988)

115. Hooker, J.D. and Jackson, B.D., 1895. *Index Kewensis.* Kew, London.

116. Howard, T.M. 1977. *Polianthes* x *Blissii* Worsley. *Plant Life,* **33** : 82-84.

117. Howard, T.M. 1978. *Polianthes* Hybrids. *Plant Life,* **34** : 126-29.

118. Hutchinson, J. 1934. *The Families of Flowering Plants.* Clarendon Press, Oxford.

119. IS 1177 : 1969. Oil of vetiver. Bureau of Indian Standards, New Delhi.

120. IS-28 : 1970. Specification for oil of Japanese mint and dementholised oil. Bureau of Indian Standards, New Delhi.

121. Indira, Raja, M., Devrajan, L. and Kanakabhushani, K. 1976. Studies on the extraction of essential oils from flowers. *Madras Agric. J,* **63** (4) : 253-54.

122. Irulappan, I., Doraipandian, A. and Muthuswamy, S. 1980. Varietal evaluation in tuberose *(Polianthes tuberosa* Linn.). *National Seminar on Production Technology for Commercial Flower Crops.* pp. 69-70.

123. Janardhan, K.K., Gupta. M.L. and Hussain, A. 1980. Effects of *Curvularia* leaf blotch disease on the essential oil content of Palma rosa. *Indian J. Exp. Biol.,* 18 (4) : 439-40.

124. Jain S.K. Medicinal Plants (India-The Land and the People), Published by National Book Trust, India (2001).

125. Jaryal G.S. and Uniyal M.R., Ausadhiya Paudho ka Vyavashaik Krishikaran (*In Hindi*), Published by Indian Society of Agri Business Professionals, New Delhi (2005).

126. Johri, A.K., Srivastava, L.J., Singh, J.M. and Rana, R.C. 1991. Effect of row spacings and nitrogen levels on flower and essential oil yields in German chamomile *(Matricaria chamomilla* L.). *Indian Perfum.,* **35** (2) : 93-96.

127. Johri, A.K., Srivastava, L.J., Singh, J.M. and Rana, R.C. 1992a. Effect of times of planting and nitrogen levels on flower and oil yields of *Matricaria chamomilla* L. *Indian J. Agron.* **37** (2) : 302-304.

128. Johri, A.K., Srivastava, L.J., Singh, J.M. and Rana, R.C. 1992b. Studies on the effect of source and time of nitrogen application in *Matricaria chamomilla. Indian J. Agric. Sci.* (under print).

129. Jana, B.K., Roy, S. and Bose, T.K. 1974. Studies on the nutrition of ornamental plants-III Effect of nutrition on growth and flowering of dahlia and tuberose. *Indian J Hortic.,* **31** : 182-85.

130. Jayaraman, V., Rajendran, G. and Muthukrishnan, T.S. 1975. Occurrence of root-knot nematodes in *Polianthes tuberosa* L. in Tamil Nadu. *Indian J Nemat.,* 5 (1) : 101-2.

131. Jindal, S.L. 1968. *Ornamental Bulbous Plants.* ICAR, New Delhi. pp. 129-30.

132. Kale, P.N. and Bhujbal, B.G. 1972. Effect of bulb on flower production of tuberose. *Indian J Orn. Hort.,* 3 102-3.

133. Kalra, A. and Parameswaran, T.N. 1988. Root-rot and wilt of *Pelargonium graveolens. Indian J. Plant Pathol.* 6 (1) : 82-83.

134. Kalra, A., Parmeswaran, T.N. and Ravindra, N.S. 1992. Influence of data on plant losses and yield response of geranium *(Plargonium graveolens)* to root rot and wilt. *J. agric Sci.* (Cambridge)118: 309-14.

135. Khan, M.N.A. 1977. Cultivation of geranium. *In Cultivation and Utilisation of Medicinal and Aromatic Plants.* (Atal, C.K. and Kapur B.M., Eds). RRL, Jammu-Tawi. pp. 316-18.

136. Korezawa, Noriaki. 1961. Studies on the development, the essential oil content and the environmental conditions in *Pelargonum* species. *Proceedings Crop Sci. Soc. Japan,* 29 (2) ; 297-300.

137. Kapoor, L.D. and Chandra, V. 1969. *Matricaria chamomilla* Linn. A new essential oil yielding crop for India. *Indian Drugs,* 6 (12) : 55-61.

138. Kapoor, L.D., Dutta, P.K. and Singh, A. 1963. Cultivation of *Matricaria chamomilla.* Regional Research Laboratory, Jammu. *Bulletin,* 1 (2) : 104-106.

139. Kaushik, T. 1987. *Effect of nitrogen and spacing on the yield of Matricaria chamomilla* Linn. M.Sc. (Forestry) thesis, Dr. Y.S. Parmar Univ. of Hort. & Fty, Solan (HP).

140. Kerches, J. 1966. Experiments on growing chemomile. Herba Jung. 5 (2-3) : 141-144.

141. Khader, Abdul, M. 1990. Perfumery concrete from fragrant flower. *Seminar on Scope of Aromatic Plants and Processing Industries,* Madurai Kamraj University, Madurai, India, 9 September 1990. pp. 1-3.

142. Krishnaswamy, N. 1990. Aromatic for industry application. *Seminaron Scope of Aromatic Plants and Processmg Industries,* 9 September 1990. Madurai Kamaraj University, Madurai, India. pp. 58-62.

143. Kumar Sushil, Srivastava RK, Singh AK, Kalra A, Tomar VKS, Bansal RP (2001) Higher yields and profits from new crop rotations permitting integration of mediculture with agriculture in the Indo-Gangetic plains. *Current Science* 80:563-566.

144. Kumari, P.R. 1979. A quarter century of research on lemongrass at L.R.S.

145. Lawrence, B.M., Hong, J.W. and Harney, P.M. 1975. Essential oils and their constituents. XIII. The chemical components of some pelargonium spp. *Int. Flav. Food Addit.,* 6 : 42-44.

146. Lawrence, B.M. 1992. Progress in Essential Oils. *Perf. Flav.,* 17 (5) : 131-46.

147. Lawrence, B.M., Powell, R.H., Smith, T.W. and Kramer, S.W.

1985. Chemical composition of *Tagetes* oil. *Perf. Flav.*, 10 (6) : 56-8.

148. Lawrence, B.M. 1984. Review of world production of essential oils. *Flavourist*, 10 : 1 20.
149. Lawrence, B.M. 1988. Progress in essential oils. *Perfume & Flavourist*, 13 : 67-76.
150. Lawless, Jaulia, Encyclopedia of Essential Oils (Element Books Ltd., Shaftesbury, Dorset, England, 1992)
151. Mabardi, M. and Khalil Mommessin, E. 1977. Some recent development in the production of jasmine concrete in Egypt. *Proc. VII Interntl. Congo Essential Oils Perfume & Flavourist*, 2 : 84.
152. Madhava Rao, V.N. 1984. Jasmine oil extraction. *Kissan World*, Feb. 20-21.
153. Maheshwari, M.L. and Sethi. K.L. 1987. Selection and improvement in palmarosa. *Indian Perfumer*, 31 (1) : 19-32.
154. Maheshwari, M.L. 1985. Gas chromatography and odour evaluation of hybrid vetiver oil. *Indian Perfumer, 29* (1-2) : 43-48.
155. Maheshwari, M.L., Srivastava, V.K., Sethi, K.L. and Gupta, R. 1986. Vetiver collection from Bharatpur, natural variability in physico-chemical constituents and odour value (pt. II). *Indian Perfumer*, 30 (4) : 47178.
156. Maheshwari, N., Krishnamurty, H.G., Chien, M.J. and Maheshwari, M.L. 1988. Composition of Jamarosa and Ocimum oils. *Indian Perfumer*, 32 (1) : 7-17.
157. Mahindra, S.N., Indian Plant Perfume (Metropolitan Book Co. Pvt. Ltd., 1, Netaji Subhas Marg, New Delhi, 1992)
158. Mani, A. K., Dakshinamoorthy, M. and Sampath, V. 1981. Effect of N, P and K on the yield and oil of geranium. *Indian Perfumer*, 25 (1) : 1-3.
159. Mani, AK, Mohan Dass, S., Kumar, N. and Sampath, V. 1981. Effect of storage time of the herbage prior to distillation on oil recovery and its quality in geranium *(Pelargonium graveolens* L. Herit). *Indian Perfumer*, 24 (3 & 4) : 35-36.
160. Mani, A.K. and Sampath, V. 1981. Seasonal influence on the oil content and quality of oil in geranium. *Indian Perfumer*, 25 (3 & 4) : 41-43.
161. Mariappan, V., Babu, K. and Kandaswamy, T.K. 1977. Leaf spot

disease of tuberose *(Polianthes tuberosa* L.) caused by a new species *Alternaria. Curro Sci.,* **46** (9) : 311.

162. Meawad, A.A., Awad, A.E. and Afity, A. 1984. The combined effect of nitrogen fertilizer and some growth regulations on chamomile plants. *Acta Horticulturae,* 144 : 123-34.
163. Misra, P.N., Chandra, V. and Kapoor, L.D. 1973. Cultivation of plants for perfumes in India-Part I. Observation on performance of tuberose *(Polianthes tuberosa* L.) on saline alkaline soils. *Indian Perfumer, 16* (1) : 38-39.
164. Misra; P.N. 1971. Effect of varying dates of planting, spacing and nitrogen levels on the growth and yield of *Matricaria chamomilla* L. Flowers in Saline Alkale Soil. *Indian Perfumer,* 15 (1) : 5-10.
165. Misra, P.N. and Kapoor, L.D. 1978. *Matricaria chamomilla* Linn. on remunerative crop for saline alkali soils. *Indian Forester,* 104 (9) : 631-37.
166. Mukhopadhyay, A and Bankar, G.J. 1981. Effect of time of planting on growth, flowering and bulb production in tuberose cv. Single. *Indian Agric.,* 25 (2) : 131-34.
167. Mukhopadhyay, A and Bankar, G.J. 1985. Effect of split application of nitrogen on growth and yield of *Polianthes tuberosa* L. cv. Single. *South Indian Hortic.,* 33 (1) : 60-62.
168. Mukhopadhyay, A and Bankar, G.J. 1986. Studies on nutritional requirement of tuberose. *South Indian Hortic.,* 34 (3) : 167-72.
169. Mukhopadhyay, A, Bankar, G.J. and Sadhu, M.K. 1986. Influence of bulb size, spacing and depth of planting on growth, flowering and bulb production of tuberose. *Haryana J. Hort. Sci.,* 15 (1 + 2) : 18-24.
170. Mookherjee, B.a., Patel, B.A., Trenkle, R.B. and Leading, W.O. 1974. Two new constituents of jasmine absolute. *Proc. Vllnt. Congo Essential Oils. Chem. Abstr.,* 85 : 5236.
171. Muthuswamy, S. and Madhava Rao, V.N. 1974. Essential oil of jasmine flowers. *Indian Perfumer,* 18 (1) : 31-34.
172. Muthuswami, S., Sambanda Murthi, S., Veluswamy, P., Japal, R. and Peter, G.B., Vedamuthu, 1980. Production economics of jasmine concrete. *Indian Perfumer,* 24 (1) : 43-45.
173. Muthuswamy, S. and Shanmugaelu, K.G. 1985. Jasmine

cultivation in South India. *In Cultivation and Utilization of Aromatic Plants* (Atal. C.K. and Kapur, B.M., Eds.). Regional Research Laboratory, Jammu. pp. 502-24.

174. Mohammad, Faiz, Nigam, M.C. and Aehman, W. 1981. Essential oil of *Cymbopogon martini* var. *mofia*. Detection of new trace constituents. *PAFAI J.* 3 (1) : 11-13.

175. Nair, E.V.G., C. Chinnamma, N.P. and Pushpa Kumari, A. 1980. Influence of spacing on the grass and oil yield of palmarosa *(Cymbopogon martini* var. motia). *Indian Perfumer,* 24 (1) : 25-27.

176. Nair, E.V.G., Chinnamma, N.P. and Kumari, P.R. 1982. Review of work done on vetiver *(Vetiveria zizanioides* Linn.) at the Lemongrass Research Station, Odakkali. In *Cultivation and Utilization of Aromatic Plants* (C.K. Atal and B.M. Kapoor, Eds.). R.R.L., Jammu. pp. 427-30.

177. Nair, E.V.G., Rajan, K.C., Chinnamma, N.P. and Kurian, A. 1983. Screening of different vetiver hybrids under Kerala conditions. *Indian Perfumer,* 27 (2) : 88-90.

178. Natarajan, S. and Madhava Rao, V.N. 1977. Studies on the influence of split application of fertilizers and farm yard manure for *Jasmine grandiflorums* L. *South Indian Hort.,* 27 (1 & 2) :111-14.

179. Natarajan, S. and Madhava Rao, V.N. 1979. Studies on response of *Jasminum grandiflorum* L. to N and P fertilization. *South Indian Hort.,* 27 (1 & 2) : 1-4

180. Natarajan, S. and Madhava Rao, V.N. 1983. Flowering and floral characters in *Jasminum grandiflorum* L. as affected by graded and split doses of fertilizers. *Indian Perfumer,* 27 (1) : 1-4.

181. Naves, Y.A. 1948. The perfume of jasmine. *Perfume. Essent. Oil Rec.,* 39 : 214-26.

182. Nambisan, K.M.P. and Krishnan, B.M. 1983. Better cultural practices for high yield of tuberose in South India.

183. Nanjan, K., Nambisan, K.M.P., Veeraragavathatham, D. and Krishnan, B.M. 1980. The effect of nitrogen, phosphorus and potash on the yield of tuberose *(Polianthes tuberosa* L.). *National Seminar on Production Technology for Commercial Flower Crops.* pp. 76-78.

184. Naragund, V.R. and Divakar, N.G. 1983. Varietal evaluation in scented geranium *(Pelargonium graveolens* L. Herit.). *Indian*

Perfumer, 27 (1) : 19-21.

185. Nair, E.V.G. 1986. Lemongrass Pamphlet NO.4. Director of Extension, Kerala Agrl. University, Trichur. Nair, E.V.G., Chinnamma, N.P. and
186. Nair, E.V.G., Chinnamma, N.P. and Nair, C.K. 1980. Optimization studies on the factors of processing of essential oils. *Indian Perfumer*, 24 (2) : 110-14.
187. Nair, E.V.G. and Nair, K.C. 1974. Studies on the cultural practices of lemon grass. *Indian Perfumer*, 18 (Part I) : 17-18.
188. Nair, E.V.G., Nair, K.C. and Chinnamma, N.P. 1979. Field experiment with micronutrients on yield of grass and oil and citral content of East Indian lemongrass, var. 00 19. *Indian Perf.*, 23 (1) : 55-58.
189. Nigam, S.S. 1982. Antimicrobial activity and essential oils. Paper presented at the Silver Jubilee Celebration and 7th Seminar of E.O.A.I. on Recent Advances in Essential Oil Research with Reference to Rural Development. 'New Delhi. 21-22 February 1982.
190. O.P., Sharma, A., Kumar, A. and Misra, L.N. 1988. Lemongrass oil (East Indian). *In Major Essential Oil Bearing Plants of India.* Central Institute of Medicinal and Aromatic Plants, Lucknow. pp. 237.
191. Ooraswamy, K.S. 1967. Curing and distillation of patchouli. *Indian Oil Soap J.*, .32 (7) : 211-16.
192. Oummond, H.M. 1960. Patchouli oil. *Perfum. Essen. Oil Record*, 51 (9) : 484-92.
193. Patil, J.D., Pati!, B.A., Chougule, B.B. and Bhat, N.R. 1987a. Effects of bulb size and spacing on stalk and flower yield in tuberose *(Polianthes tuberosa* L.) cv Single. *Current Research Report.* MPAU, Maharashtra 3 (2) : 81-82.
194. Pareek, S.K. and Gupta, R. 1985. On the status of agronomic research in *Cymbopogon* grasses in India with projections on future work. *Indian Perfumer*, 29 (3&4) : '215-24.
195. Pillai, K.P. 1961. Lemongrass in India. Farm Bull. No. 16. Ministry of Food and Agriculture, New Delhi. pp 26.
196. Pareek, S.K. and Gupta, R. 1985. On the status of agronomic research in *Cymbopogon* grasses in India with projection on future

work. *Indian Perfumer,* 29 (3&4) : 215-24.

197. Pareek, S.K., Maheshwari, M.L. and Gupta, R. 1981 a. Cultivation of palmarosa oil grass. *Indian Fmg.,* **36** (4) : 37-40.

198. Pareek, S.K., Maheshwari, M.L. and Gupta, R. 1981 c. Effect of N, P and K fertilizers on the yield of palmarosa grass under cultivation. *Indian J. Agron.,* **26** (2) : 122-29.

199. Pareek, S.K., Maheshwari, M.L. and Gupta, R. 1983a.- Effect of density of planting on the yield of palmarosa. *Indian J. Agric. Sci.,* **53** (6) : 466-77.

200. Pareek, S.K., Maheshwari, M.L. and Gupta, R. 1984. Effect of FYM and micronutrients on yield and quality of palmarosa grass. *Indian Perfumer,* **28** (2) : 108-11.

201. Pareek, S.K., Maheshwari, .M.L. and Gupta, R. 1985. Studies on palmarosa oil grass germplasm, Part III. Evaluation of high yielding genotypes at Delhi. *Indian Perfumer,* **29** (3&4) : 209-14.

202. Pareek, S.K., Maheshwari, M.L. and Gupta, R. 1991. Chemical weed control and water management in periwinkle, vetiver and palmarosa oil grass. *Proc. Intern. Seminar on Medicinal Aromatic and Spice Crops.* Vol. 1 (S.P. Raychaudhari, Ed.). Today & Tomorrow Printers & Publishers, New Delhi. pp. 167-70 .

203. Pareek, S.K., Maheshwari, M.L., Singh, K.D. and Gupta, R. 1983b. Nutrient uptake and dry matter production of palmarosa oil grass under different levels of N, P and K fertilizers. *Intern J. Trop. Agric,* 1 (3) : 203-09.

204. Pareek, S.K., Mandai, S., Maheshwari, M.L. and Gupta, R. 1980.

205. Pareek, S.K., Saxena, D.B., Maheshwari, M.L. and Gupta, R. 1981b. Study of palmarosa oil grass for higher yield of oil and its quality under cultivation. *Indian Perfumer,* **25** (3&4) : 64-70.

206. Pareek, S.K., Maheshwari, M.L. and Gupta, R. 1991a. Intercropping in vetiver *(Vetiveria zizanioids L.).'Indian Perfumer,* 35 (4) : 235-38.

207. Pareek, S.K., Maheshwari, M.L. and Gupta, R. 1991c. Recent advances for improving vetiver *(Vetiveria zizanidoes)* cultivation in India. *Symp. on New Trends in Essential Oils.* Tata Mc. Graw-Hill Publishing Co. Ltd. pp. 269-79.

208. Pareek, S.K., Maheshwari, M.L. and Gupta, R. 1992. It pays to grow 'hybrid-8' vetiver. *Indian Fmg.,* 41 (12) : 8-10 .

209. Prajapati N.D., Jadi Butio Ki Saral Kheti (*In Hindi*) Published by Rajasthan Agro-Forestry Corporation, Jodhpur (2002)

210. Polak, H.E. 1973. Recent progress in jasmine research. *Cosmo Perf.,* 88 : 46-48.

211. Punia, M.S., Verma, P.K. and Sharma, G.D. 1989. Heritability estimates and performance of vetiver *(Vetiveria zizanioides)* hybrids. *Indian J. Agric. Sci.,* 59 (1) : 71-72.

212. Pushpangadan, P., Sobti, S.N. and Khan, R. 1975. Inheritance of major essential oil constituents in *Ocimum bacilicum* var. *glabratum* Benth. (French basil). *Indian J. Exper. Bioi.,* 13 (5) : 520.

213. Prakasa Rao, E.V.S., Murinu Singh, Narayana, M.R. and Chandrasekhara, G. 1985. Effect of N, P and K nutrition on herb and oil yield of geranium. *Indian Perfumer,* 29 (3-4) : 1'47-50.

214. Penka, M. 1967. The effect of irrigation on the water economy and essential oil accumulation in *Matricaria chamomilla. Acta. University Agric. Fac. Agronom.,* 15 : 527-39.

215. Ramanujam, S. and Sushil Kumar. 1963a. Correlation studies in two populations *of Vetiveria. IndianJ. Genet.,* 23 (1) : 82-89.

216. Ramanujam, S. and Sushil Kumar. 1963b. Multiple criteria selection in vetiver. *Indian J. Genet.,* 23 (2) : 17684.

217. Razdan, T.K., Wanchoo, R.K. and Dhar, K.L. 1986. Chemical composition and antimicrobial activity of the oil of *Tagetes minuta* Linn. *Perfum. Kosmet.,* 67 : 52-55.

218. Reichling, J.W:, Bisson, W. and Becker, H. 1984. Comparative study on the production and accumulation of essential oil in the whole plant and in the callus culture of *Matricaria chamomilla. Planta Medica,* 51 (4) : 324-39.

219. Repeak, M. Halasova, J., Honcariv, R. and Podhradsky, D. 1980. The content and composition of the essential oil in the course of quthodium development in wild chamomile *(Matricaria chamomilla* L.). *Biologia plantanum,* 22 (3) : 183-291. (Czechoslovakia).

220. Ramaswami, N. and Chockalingam, P. 1977. Influence of weight of tubers on yield of tuberose *(Polianthes tuberosa* L.). *Progressive Horticulture,* 8 (4) : 39-41.

221. Roy, A.K. 1985. Fungal diseases of ornamentals and flowers in Assam. *Indian Hortic.,* 30 : 19-21.

222. Rajeswara Rao, B.R., Prakasa Rao, E.V.S. and Narayana, M.R. 1989. Rose geranium, an economical crop in the south Indian plains. *Indian Hortic.*, 34 (2) : 14-17.

223. Rakesh Tewari, and Virmani, O.P. 1988. Chemical studies in geranium oil: A review. *CROMAP*, 10 (2) : 72-81.

224. Ranade, G.S. 1988. Chemistry of geranium oil. *Indian Perfumer*, **32** (1) : 61-68.

225. Randhawa, G.S., Mahey, R.K., Sidhu, B.S. and Saini. 5.5. 1984a. Effect of nitrogen application on herb and oil yield of *Mentha citrata*. *Indian Perfum.*, 28 (2) : 105-7.

226. Sandhu, G.S., Singh, B. and Bhalla, J.S. 1975. Insect pests of Japanese mint in India. *Indian Perfumer*, 19 (1) : 13-18.

227. Sadhu, M.K. and Das, P.C. 1978. Effect of bulb size, planting density and depth of planting on growth, flowering and bulb production of tuberose *(Polianthes tuberosa* Linn). *Indian J. Hortic.*, 35 : 147-50.

228. Sane, P .V. 1986. Floriculture research at NBRI. *Ornamental Horticulture in India* (KL Chadha and B. Choudhury, Eds.). pp. 20-37 .

229. Sharga, AN. 1977. Tuberose for elegance and fragrance. *Indian Hortic.*, 21 : 25-29.

230. Sharga, A.N. 1980. Maximizing tuberose *(Polianthes tuberosa* Linn.) multiplication by using bulb segments. *Progressive Horticulture*, 12 (2) : 71-77.

231. Sharga, AN. and Motial, V.S. 1978. Studies on the effects of foliar nutrition on the vegetative growth, floral characters and essential oil contents of tuberose *(Polianthes tube rosa)* I. Vegetative characters. *Indian J. Hortic.*, 35 (3) : 244-53.

232. Sarwar, M. 1973. The impact of *Fusarium* wilt on the production of geranium oil. *Indian Perfumer*, 17 : 5254.

233. Sarwar, M., Parameswaran, T.N., Shanmugam, C. and Narayana, M.R. 1982. Chemical and cultural management of wilt of geranium. *Indian Perfumer*, 26 (2-4) : 122-24.

234. Satchith, K.M. and Sundaram, M. 1987. Production of essential oil and problems facing essential oil industry. *Indian Pefumer*, **31** (1) : 41-43.

235. Sharma. J.S., Bains, J.S. and Gilt. G.S. 1971. Nitrogen and phosphorus uptake of *Mentha arvensis* L. *Indian Perfum.*, 19 (2) :

17-20.

236. Singh, V.P.: Duhan, S.P.S. and Roy, S.K. 1978. Cropping pattern studies with Japanese mint *(Mentha arvensis* L.) in Tarai of Nainital. *Indian Perfumer,* 22 (3) : 107-9.

237. Singh N., Luthra, R. and Sangwan, R.S. 1989. Effect of leaf position and age on the essential oil quantity and quality in lemongrass. *Planta. Medica.,* 55 : 127-33.

238. Singh, R.S., Pathak, M.G. and Bordoloi, D.N. 1983. Citral of lemongrass under different plant population. *Pafai. J.,* 5 (1) : 33-36.

239. Sethi, K.L., Maheshwari, M.L. and Gupta, R. 1986. Genetic resources of palmarosa oil grass and their utilization. *Indian Perfumer,* **30** (2-3) : 309-13.

240. Sharma, J.S., Saini., S.S. and Pains, D.S. 1977. Influence of row spacing and N and P application on the fresh herb and oil yield of palmarosa *(Cymbopogon martini* var. *motia). Indian Perfumer,* **21** (1) : 44-46.

241. Sharma, S.N., Singh, A. and Tripathi, R.S. 1980. Response of palmarosa to nitrogen, phosphorus and zinc. I*ndian J. Agron.,* **25** (4) : 719-23.

242. Shukla, S.V. & Dwivedi, K.N., Utilization of Bio-resources (Proceedings of the National Conference Utilization of Bio-resources, October 24 – 25, 2002) Regional Research Laboratory, Bhubaneswar page 75-82.

243. Surburg, H. 1988. Minor components of palmarosa oil. *Development in Food Science in Flavour and fragrances.* A *World Perspective* (B.M. Lawrence, B.D. Mookerjee and B.J. Wills, Eds.). Elsevier, Amsterdam. pp. 505-15.

244. Sethi, K.L. 1982. Breeding and cultivation of new *khas* hybrid clones. *Indian Perfumer,* 26 (2-4) : 54-61.

245. Sethi, K.L., Gupta, R. and Mehra, R.B. 1981. Stability analysis of root yield and oil content in vetiver. *Indian J. Agric. Sci.,* 51 (5) : 299-301.

246. Sethi, K.L., Maheshwari, M.L. and Gupta, R. 1992. Exploiting natural variability in Bharatpur vetiver. *Symp. Proc. Recent Advances in Medicinal, Aromatic and Spice Crops.* Today and Tomorrow Printers and Publishers. Delhi. Vol. 2, pp. 457-61.

247. Sethi, K.L., Maheshwari, M.L., Srivastava, V.K. and Gupta, R. 1986. Natural variability in *Vetiveria zizanioides* collections from Bharatpur (pt. I). *Indian Perfumer*, 30 (2-3) : 377-80.

248. Singh. A . Singh, M. and Singh, D.V. 1987. Vetiver-A promising crop for problem soils. *Pafai J.*, 9 (2) : 2729 .

249. Singh, Iqbal, Singh. J.P. and Chaturvedi, O.P. 1978. Studies on essential oil of *khus (Vetiveria zizanioides* Stapt). *Indian Perfumer* 22 (3) : 218-20.

250. Singh, B., Sood, R.P. and Singh, V.1992. Chemical composition of *Tagetes minuta* L. oil from Himachal Pradesh (India). *J. Ess. Oil Res.*, 4 : 525-26.

251. Singh AK, Srivastava RK, Tomar VKS, Kumar S. (1997). Some important aromatic crops and their distillation system. (in Hindi). *Udamita Samachar (MP)*; Sep. 1997. p-(15-18).

252. Shah, N.C. 1980. Traditional cultivation of 'Babunah' Chamomilla recutita (L.) Raushert Syn. *Matricaria chamomi/la* L. in Lucknow. *Bulletin Medico-Ethnobotanical Research*, 1 (3) : 471-77.

253. Singh, A. 1969. Cultivation of *Matricaria chamomilla*. Paper presented *Symposium on Raising of Medicinal Herbs*. RRL, Jammu. March 1969.

254. Singh, A. 1982. Cultivation of *Matricaria chamomilla*. In cultivation and utilization of Aromatic Plants (Atal, C.K. and Kapoor,' B.M., Eds.). Regional Research Laboratory, Jammu, pp. 653-58.

255. Singh AK, Srivastava RK, Kalra A, Bansal RP, Tomar VKS and Kumar S. (2003). New Practices in Cultivation of the Mint, *Mentha arvensis* in the Indo-Gangetic plains. *Experimental Agriculture (Cambridge University Press)* UK 39: 199-207.

256. Singh AK, Srivastava RK, Kumar S. (1999); Production and trade of menthol. *FAFAI J*. April – June, 1999-(28 –32).

257. Singh J, Srivastava RK, Singh AK, Kumar S. (2000); Current status of production and trade of major essential oils. *J. Med. Aro. Pl. Sci.* 22 (723-731).

258. Singh J, Bagchi GD, Srivastava RK, Singh AK (2000); Pharmaceutical aspects of aromatic herbs and their aroma chemicals. *J. Med. Aro. Pl. Sci.* 22 (732-738).

259. Singh J, Srivastava RK, Singh AK, Kumar S.(2001). Futuristic scenario in production and trade of major medicinal plants of

India. *J. Med. Aro. Pl. Sci.* 22 & 23 : (564-571).

260. Singh, S.P., Sastry, K.P. and Sharma, S. 1989. Effect of growth regulators on rooting and yield of geranium. *pA FA I, J.*, **11** (2) : 35-36.

261. Sharma, M.L., Singh, A. and Juneja, B.R. 1980. *Extraction of jasmine Flower Oil.* Extension Bulletin. No. 7. National Botanical Research Institute, Lucknow.

262. Srivastava, H.K. 1988. Phenotypic patterns in relation to herbage and oil yield in *Cymbopogon martini Wats. var. motia . Int. J. Crude Drug Res.*, **26** (1) : 22-28.

263. Srivastava, H.C. 1987. Annual Report. Indian Institute of Horticultural Research, Bangalore.

264. Srivastava, H.C. 1987. Studies on possibilities of commercial exploitation of close media sachet technique for propagation of jasmines (Hindi language). *Bhartiya Krishi Anusandhan Patrika*, 2 (1) : 47-52.

265. Srivastava, H.C. and Sharma, M.L. 1971.. Cultivation of *Jasminum grandiflorum* L. on saline soils. *Indian Perfumer*, **18** (1) : 25-26.

266. Srivastava RK, Singh AK, Kalra A, Bansal RP, Tomar VKS, Patra DD, Naqvi AA, Chand S, Sharma S and Kumar Sushil. (2002). Characteristics of menthol mint *Mentha arvensis* L. crop cultivated on industrial scale in the Indo- Gangetic plains, *Industrial Crops and Products*, USA 15:189-198.

267. Srivastava RK, Singh AK, Bansal RP, Kalra A, Bahal JR, Tomar VKS, Patra NK, Kumar S. (2000); Adoption of improved varieties of menthol mint (*Mentha arvensis*) Himalaya and Kosi in Uttar Pradesh. *J. Med. Aro. Pl. Sci.* 22 *(470-473).*

268. Srivastava RK, Singh AK, Kalra A, Bansal RP, Tomar VKS, Bahl JR, Sharma S and Kumar Sushil. (2000). Optimum crop age of menthol mint *Mentha arvensis* cv. Kosi crops in the Indo-Gangetic plains for high yields of menthol rich essential oil. J. Med. Aro. Pl. Sci. 22 (771-773).

269. Srivastava RK, Singh AK, Kalra A, Tomar VKS, Bansal RP and Kumar S. (2003). Influence of age and literacy level of farmers on adoption of mint based crop rotations in the Indo-Gangetic plains. J. Med. Aro. Pl. Sci. 25 689-697.

270. Srivastava RK, Lehri A. and Singh DP. (2005). Integration of aroma bearing crops with agriculture in tribal areas of Gujarat

for uplifting rural economy – A study. FAFAI Journal. Jan-Mar (63-69).

271. Srivastava, RK. (2001). Low productivity of Indian agriculture: reasons & remedies. (*In Hindi*). CIMAP Magazine *Sugandha* 2 : (22-24).

272. Srivastava, RK (2005) Improved cultivation of Menthol mint (*in Hindi*). Udamita Samachar 3: 29-31.

273. Srivastava, RK (2005) Unutilized land may be utilized by cultivation of scented Khus/Vetiver (*in Hindi*). Udamita Samachar 8:14: 77-79.

274. Srivastava, RK (2006) Cultivation of German chamomile for blue oil (*in Hindi*). Udamita Samachar 2:15: 7-10.

275. Srivastava, RK (2006) Value addition in Essential oils (*e-magazine*). Science Tech- Entrepreneurs April, 2006.

276. Srivastava, RK (2006) Value addition in Essential oils (*In Hindi*). Udamita Samachar 6/2006: 52-56.

277. Srivastava, RK (2006) More income with less investment (*In Hindi*). Udamita Samachar 15/7/2006: 28-30.

278. Srivastava, RK (2006) Improved cultivation of Palmarosa (*In Hindi*). Udamita Samachar Nov. 2006: 29-30.

279. Srivastava RK and Raghav CS (2006). Emerging trends in cultivation and processing of aromatic crops in tribal areas of Gujarat for boosting-up their socio-economic condition. *Indian Perfumer*.

280. Swaminathan, K.R., Muthuswamy, S. and *Madhava* Rao, V.N. 1979. Pilot plant for extraction of jasmine essential oil from *Jasminum grandiflorum*. *Indian Hort.*, **24** (1) : 20-21.

281. Thapa, R.K., Agarwal, S.G., Kalia, N.K. and Kapoor, R. 1993. Changes in chemical composition of *Tagetes minuta* oil at various stages of flowering and fruiting. *J. Ess. Oil Res.*, 5 : 375-79.

282. Tiwari, Rakesh, and Virmani, D.P. 1988. Chemical studies in geranium oil : A review. *CROMAP*, **10** (2) : 72-81

283. Timmer, R., Heide, R., Valios, P.J. and Wobben, H.J. 1971. Qualitative analysis of the most volatile neutral components of Reunion geranium oil. *J. Agric. Food Chem.*, **19** ; 1066-68.

284. Thimmappa, O.K., Bhattacharjee, S.K. 1990. Standardisation of propagation of scented geranium from stem cuttings. *Indian*

Perfumer, **34** (1) : 56-60.

285. Thomas, J., Joy, P.P. and Geetha, K. 1989. Optimum method of planting in lemongrass. *Indian Perfumer*, 33 (2) : 102-103.
286. Tyagi, NK and Srivastava, RK. (2001). Agricultural pollution and affects on environment. (*In Hindi*). CIMAP Magazine *Sugandha* 2 : 41-43.
287. Urinson, R.P. 1936. Studies on the chemical composition of the essential oils of mutants of the *Pelargonium roseum* Hort. *Bull. Appl. Bot. Leningr. Ser.*, III (13) : 67-85 ..
288. Vadivel, B. and Sampath, V. 1981. Evaluation of promising clones of geranium *(Pelargonium graveolens* L. Herit.). *IV Workshop on All India Coordinated Project on Medical and Aromatic Plants.* 31 August 3 September 1981. Madurai.
289. Varshney, S.C. 1993a. Efficacy of steam distillation units for herbs. *Indian Perfumer*, 37 (1): 63-67.
290. Virmani, O.P., Srivastava, A. and Srivastava, G.N. 1977. Lemongrass and its Cultivation in India. Farm Bull. No.6, CIMAP, Lucknow. pp. 14.
291. Virdi M.S., Malvia S., Aromatic Plants & Essential Oils (2007).
292. Wali, A. 1980. Yield and volatile oil content of chamomile *(Matricaria chamomilla* L.) as affected by water supply *Herba Hung.*, 19 (3) : 65-72.
293. Wanat, Krogulska, A. 1952. Effect of various nutritive ingredients on the yield and quality of *Matricaria chamomilla Desst. Pharm.*, 4: 257-75.
294. Yadav, L.P., Bose, T.K. and Maity, R.G. 1984. Effect of bulb size and depth of planting on growth and flowering of Tuberose *(Polianthes tuberosa* Linn). *Progressive Horticulture*, 16 (3 & 4): 209-13.
295. Yadav, L.P., Bose, T.K. and Maity, R.G. 1985. Response of tuberose *(Polianthes tuberosa* L.) to nitrogen and phosphorus fertilization. *Progressive Horticulture*, 17 (2): 83-86.
296. Yoshida, T. 1969. On characteristics and some problems in cultivation of *Pelargonium* species. *Japan Agric. Res. Quart.*, 4 (3): 23-26.

Index

Plate 1 : A field view of Sweet Basil

Plate 2 : A Close-up view of Citronella

Plate 3 : A field view of Chamomile at flowering stage

Plate 4 : A Close-up view of Geranium

Plate 5 : A Close-up view of Jasmine plant

Plate 6 : A field view of Lavender

Plate 7 : A Close-up view of Lemongrass

Plate 8 : A Close-up view of Menthol mint

Plate 9 : A flowering plant of Marigold

Plate 10 : A view of mature plant of Palmarosa

Plate 11 : A matured plant of Patchouli

Plate 12 : A flowering plant of Damask Rose

Plate 13 : A field view of Rosemary

Plate 14 : Flowering twig of Tuberose

Plate 15 : A close-up view of Vetiver Plant

Plate 16 : A view of Hydro-distillation (Deg-Bhapka Method)

Plate 17 : A view of Hydro-Steam Distillation (Field Distillation Unit)

Plate 18 : A view of Steam Distillation Unit